T0074864

Essentials of Mobile Handset Design

Discover what is involved in designing the world's most popular and advanced consumer product to date – the phone in your pocket. With this essential guide you will learn how the dynamics of the market, and the pace of technology innovation, constantly create new opportunities which design teams utilize to develop new products that delight and surprise us. Explore core technology building blocks, such as chipsets and software components, and see how these components are built together through the design lifecycle to create unique handset designs.

Learn key design principles to reduce design time and cost, and best practice guidelines to maximize opportunities to create a successful product. A range of real-world case studies are included to illustrate key insights. Finally, emerging trends in the handset industry are identified, and the global impact those trends could have on future devices is discussed.

ABHI NAHA is the CEO and founder of Zone V Ltd, which specializes in empowering blind and partially sighted people through mobile devices. Previously he held senior management roles at Powermat and Idem. Prior to that he was a senior manager at a Silicon Valley mobile touchscreen and user interface technology provider, Synaptics. He is an Ex-Chairman of the UK-based charity "Beatbullying" and Vice President and board member of the Communications and Manufacturing Association of India. He has also held advisory roles for Goldman Sachs and Silver Lake in the area of mobile handset user interface technologies. He holds an MBA from Aston Business School and a B.Sc. (Hons.) degree in Electronics Engineering from Leicester University.

PETER WHALE is Director of Product Management for the Qualcomm subsidiary, Xiam Technologies. During his career in the mobile industry he has created numerous innovative technology products shipped in handsets in very high volumes and built into network services used by consumers many times a day. At Qualcomm, he leads the conceiving and development of innovative software products, applying machine learning techniques to the fields of device optimization, device personalization and content discovery. He is a board member of Cambridge Wireless, where he leads the team who deliver the annual "Future of Wireless" International Conference. He holds a degree in Computer Science, is a member of the British Computer Society, and is a Chartered Engineer.

The Cambridge Wireless Essentials Series

Series Editors
WILLIAM WEBB, *Neul, UK*
SUDHIR DIXIT, *HP Labs, India*

A series of concise, practical guides for wireless industry professionals.

Martin Cave, Chris Doyle and William Webb, *Essentials of Modern Spectrum Management*
Christopher Haslett, *Essentials of Radio Wave Propagation*
Stephen Wood and Roberto Aiello, *Essentials of UWB*
Christopher Cox, *Essentials of UMTS*
Steve Methley, *Essentials of Wireless Mesh Networking*
Linda Doyle, *Essentials of Cognitive Radio*
Nick Hunn, *Essentials of Short-Range Wireless*
Amitava Ghosh and Rapeepat Ratasuk, *Essentials of LTE and LTE-A*
Abhi Naha and Peter Whale, *Essentials of Mobile Handset Design*

Forthcoming
David Bartlett, *Essentials of Positioning and Location Technology*

For further information on any of these titles, the series itself and ordering information see www.cambridge.org/wirelessessentials

Essentials of Mobile Handset Design

Abhi Naha
Zone V Ltd

Peter Whale
Qualcomm

Shaftesbury Road, Cambridge CB2 8EA, United Kingdom

One Liberty Plaza, 20th Floor, New York, NY 10006, USA

477 Williamstown Road, Port Melbourne, VIC 3207, Australia

314–321, 3rd Floor, Plot 3, Splendor Forum, Jasola District Centre, New Delhi – 110025, India

103 Penang Road, #05–06/07, Visioncrest Commercial, Singapore 238467

Cambridge University Press is part of Cambridge University Press & Assessment, a department of the University of Cambridge.

We share the University's mission to contribute to society through the pursuit of education, learning and research at the highest international levels of excellence.

www.cambridge.org
Information on this title: www.cambridge.org/9781107010048

First published 2012

A catalogue record for this publication is available from the British Library

Library of Congress Cataloging-in-Publication data
Naha, Abhi, 1966–
Essentials of mobile handset design / Abhi Naha, Peter Whale.
pages cm. – (The Cambridge wireless essentials series)
ISBN 978-1-107-01004-8 (hardback)
1. Cell phones – Design and construction. I. Whale, Peter. II. Title.
TK6564.4.C45N34 2012
621.3845′6 – dc23 2012021824

ISBN 978-1-107-01004-8 Hardback

I would like to thank my friend Cherie Blair for her support and encouragement in using mobile technology to empower humanity and my mother, Lovely Naha, who taught me the power of education. Special thanks to Sammy, Jy and Shaan for their love.

Abhi Naha

To my loving family – Janice, Ben, Cathy and Sam, and to my mother, Janet Whale, who taught me the power of goals.

Peter Whale

"The mobile phone has transformed the world. Not only has it delighted consumers in the West, but it has been at the heart of a bottom up innovation revolution in Africa, Asia, and Latin America. Naha and Whale, in their various roles throughout their varied careers, have been a central part of this revolution and will continue to shape it in the future. This book draws on their rich experience. It is a remarkable resource for all (and there are many) who have an interest in mobile handset technology."

Jaideep Prabhu, Cambridge Judge Business School

"Mobile handset design is incredibly important, affecting the lives of most of us, and the fate of major corporations. This year there will be more handsets than people on the planet. Handsets are one of the most ubiquitous artefacts made by man, yet their inner workings are magical and mysterious to many. Starting from a historical perspective our two wizards take the reader step by step through an overview of the internals and the design process before concluding with a chapter on future trends, which alone is worth buying the book to read. Essential for anyone in or interested in the mobile industry."

Jack Lang, Serial Entrepreneur and Fellow of University of Cambridge

"As an interested amateur in the mobile world I found it fascinating to learn of the history and evolution of this simple but powerful tool which has transformed our world."

Cherie Blair, Founder of Cherie Blair Foundation for Women

"Well done! It's highly informative and incredibly detailed. Design of mobile devices requires a complex and seemingly endless series of deeply varied and dynamic considerations. Few people can appreciate the incredible team effort required to create a truly winning solution. This book shares the multitude of layered technologies that have been at the core of my evolving world as a mobile design leader."

Frank Nuovo, Design Studio Nuovo, Former Chief of Design for Nokia and Vertu

"Mobile phones now have 500,000 times as many transistors as the first working device in the early 70's. But as this book shows, it is possible to understand the whole process of designing one of these extraordinary machines. Taking an 'holistic' approach, Abhi Naha and Peter Whale touch on all aspects of mobile phone design, whether it be branding, software, electronics, functionality or project management, and they show that good design requires thinking about the issues from many different points of view.

"We are entering a new era of open hardware and software, where new gadgets will be brought into existence by groups whose members are distributed around the world. This book sets out not only how to design one of the 6 billion phones on the planet today but also points the way as to how many will be designed in the future. A must read for anyone interested in mobile handset design."

David Cleevely, Serial Entrepreneur and Chairman of Cambridge Wireless

"The book describes in a clear and concise way the extraordinary progress in handset design over the last two decades. If as a student or practising engineer you need to know how one of the most important components of the global digital infrastructure is engineered then I would recommend this book. Some of the most popular handsets are described and put in context. The crucial role of design is emphasised and future trends and development paths are enumerated. An excellent read about an indispensable piece of equipment for everyone"

Andy Hopper, Head of Department, The Computer Laboratory, University of Cambridge

"Tens of thousands of mobile handsets have been launched; yet mere dozens have become spectacular successes. Each triumphant handset has shared one key attribute – good design, inside and out. Good engineering matters little if its functionality is rendered inaccessible by flawed interface design. This book is a useful, comprehensive guide to the myriad factors that collectively create a winning design."

Paul Lee, Director / Technology, Media & Telecommunications, Deloitte LLP

Contents

	Preface	*page* xi
	Acknowledgements	xiv
1	**Beginnings**	1
	1.1 Development of the first mobile handset	1
	1.2 Generations of mobile communication capability	8
	1.3 The digital revolution – 2G	9
	1.4 High-speed data – 3G	30
	1.5 Mobile broadband – 4G	31
	1.6 Conclusion	31
	1.7 Timeline of the mobile phone	32
2	**Design influences**	35
	2.1 Core design influence – engagement value	35
	2.2 Trends	37
	2.3 Desirability	39
	2.4 User journey	41
	2.5 Usability	43
	2.6 Corporate user versus personal user	46
	2.7 Business models	46
	2.8 New entrants into the mobile industry	48
	2.9 Content	49
	2.10 Applications and services	51
	2.11 Internet	52
	2.12 Innovation in handset design	52
	2.13 Legislation	53
	2.14 Conclusion	54

3 **Design architecture** 56
3.1 Design perspective 56
3.2 Physical view – a product teardown 58
3.3 Standards view 67
3.4 Hardware architecture view 70
3.5 Software architecture view 77
3.6 Manufacturer view 83
3.7 Operator view 86
3.8 Summary 88

4 **Hardware design** 90
4.1 Introduction 90
4.2 Helicopter view 90
4.3 The radio spectrum 93
4.4 Radio chipset design 96
4.5 Digital chipset design 108
4.6 Baseband cellular modem design 117
4.7 Mobile application processor design 123
4.8 Multimedia processor design 126
4.9 Peripheral component design 128
4.10 Conclusion 129

5 **Software design** 130
5.1 Application software design 131
5.2 Protocol stack software design 145
5.3 Physical layer software design 153
5.4 Mobile operating systems and execution environments 156
5.5 Conclusion 161

6 **Product design** 162
6.1 Introduction 162
6.2 The design process 163
6.3 Planning a handset design program 166
6.4 Brand DNA design 171
6.5 Visual and desirable design 173

6.6 User experience design 174
6.7 Industrial design 177
6.8 Mechanical design engineering 180
6.9 Hardware design engineering 183
6.10 Software platform design 184
6.11 Manufacturing production 186
6.12 Testing and qualification 187
6.13 Channel marketing design 187
6.14 Case study: capacitive touchscreens in mobile handsets 188
6.15 Using the design process as a strategic tool for innovation 191
6.16 Inclusive design and accessibility 192
6.17 Conclusion 193

7 **Future trends** 195
7.1 Introduction 195
7.2 The journey 196
7.3 Designing for the environmental footprint 199
7.4 The "smart" journey 201
7.5 The mobile home 205
7.6 Near-field sensing 206
7.7 Augmented reality 208
7.8 Wireless power charging 209
7.9 Case study: wireless charging in smart devices 210
7.10 Social values and shared experiences 213
7.11 Efficient handset platforms 219
7.12 Handsets and the retail experience 221
7.13 Summary 222
7.14 Conclusion 223

8 **Conclusion** 226

Appendix: User interaction and experience design phases 229
Glossary 231
Index 235

Preface

On October 31, 2011, statisticians at the United Nations believe that the world reached a global population of seven billion people. Industry analysts estimate that, a few weeks later, the six billionth active cellular phone connection was achieved. Even considering that some people have multiple phone connections, it is still incredible that the majority of human beings on our planet own a mobile handset – across all geographies, cultures and societies. The ability to connect to people and information, wherever you are and whenever you want to, is bringing into reality the concept of the global village and a shared humanity. Yet at least as amazing is the fact that this ubiquitous product is also the world's most advanced consumer electronics product ever. It is small, light, portable and affordable to most people. It can perform ever more functions, with increasing performance. How is this possible and how has this come to be? What really is involved in designing and bringing to market both the inner technology and the final desirable mobile handsets? How can new products be developed so quickly, and why are there so many to choose between?

This book is aimed at anyone who is curious about such questions, yet has limited time to invest. You can read the book completely over a weekend if you need a lot of knowledge quickly, or you can dip in and out of the book to gain insight into particular areas as you need them. Blaise Pascal in his *Provincial Letters* wrote "I would have written a shorter letter, but I did not have the time." We have found the truth in this, as condensing down the key insights from as complex an industry as the mobile handset industry into the convenient format of the *Wireless Essentials* series has been a significant undertaking – it would have been easier to write a larger volume. However, the format offers you the benefit of being able to gain a really good appreciation of the technology, design processes and market issues involved in mobile handset design relatively quickly. As you read the book, we trust that you will gain good insight into a set of common design and market issues which re-appear consistently as we look at different aspects of both the underlying component technology and the handset product design process.

We are both practitioners in the mobile industry. We have spent most of our working lives living and breathing the topics we cover, and much of what we write is from first-hand experience. Our approach is to introduce you to the fundamental design and engineering principles used in creating handsets and their component technologies, and to give you an "under the hood" understanding of a mobile handset. We have sought to do this in such a way so that whether you are a curious consumer, a technologist, designer, developer, lecturer, student or brand marketer, you will gain a good appreciation and enough understanding to hold useful discussions with other practitioners in the mobile industry – or enjoy using your mobile handset more because of a deeper appreciation of its inner workings and the process of creating it.

One of the takeaways of this book is that such is the complexity of the technology, the industry and the global market, that no one can claim to be truly an expert of it all. Rather, it is by many thousands of talented designers, engineers, marketers and others bringing their expertise together and working well together as teams, organizations, industries and ecosystems that the phenomenon which is the mobile handset is possible. To do this well requires an understanding and appreciation of the design challenges between different disciplines, and the ability to take a holistic view of the whole design process and value chain. Our aim therefore is not to make you an expert, but rather to raise your general level of knowledge and understanding of what is involved in the design of a mobile handset, as well as the level of communication and partnerships needed to achieve success. If you work in any aspect of the mobile industry, we hope that this book will help you to reach out beyond your domain expertise. Our aim is that, with the extra knowledge and insight gained from reading this book, you will benefit from dialog and partnership with experts from other domains, and that this will lead to you achieving your business goals more rapidly and successfully.

Regarding the use of the term "mobile handset," we have had several discussions on using the term "handset" as opposed to any other term, such as mobile phone, cell phone or connected device, particularly as new classes of portable devices enter the market such as tablets, e-readers and the like. One piece of future-oriented research suggests that mobile handsets will take the form of tiny implants so perhaps we should be using the term "headset" rather than "handset." For simplicity and consistency, we settled on using the term "mobile handset" throughout the book to refer to all types of cellular device, irrespective of form-factor, market or geography.

In writing a book about an industry which continues to develop at such a fast pace, one area of concern is about keeping any book about the mobile handset up to date. Our approach has been to describe the fundamental handset design principles and strategies that we believe have remained constant in the journey so far, and perhaps will remain so for many years to come. What undoubtedly will change is the creation of new technologies, applications, services and markets, as the fuel of the mobile handset industry is continued and rapid innovation. The principles we describe should help you to continue to evaluate the benefits of further innovation and the value they add to the design process in creating engaging and desirable mobile handset user experiences. Lastly, with the agreement of our publishers, we have created the website www.mobilehandsetdesign.com to capture some of the new trends, debates and discussions which will unfold in the future, and we will aim to keep this fresh and up to date. Finally, we hope you enjoy reading this book as much as we have enjoyed writing it.

Acknowledgements

Abhi would like to thank Frank Nuovo and Dr Peter Ashall for the very lively and passionate discussions on mobile handset design, and Ken Blakeslee for keeping Abhi grounded on the importance of end-user value.

Peter would like to thank Gary Dibley, Pascal Herczog and Paul Tindall for their invaluable input and comments on early drafts of his chapters, which ensured he was setting off down the right road. A word of thanks also to all the staff at the coffee shop within the supermarket in Milton near Cambridge, where most of Peter's writing was done – for their early morning smiles and good humor, which always gave him encouragement.

A special thank you from both of us to Sarah, Julie, Lucy, Elizabeth, Irene and the team at Cambridge University Press for their forbearance, patience and encouragement during this writing project (it took somewhat longer than we all expected), and to Professor William Webb for his valuable guidance and support in keeping us on track with our book.

1 Beginnings

The mobile phone industry is one which has been characterized by a breathtaking speed of change and development, and anyone who has owned a number of handsets will be aware of the dramatic change evident between a phone of just a few years ago and the latest available models. In order to identify a set of core design issues which hold across generations of handset design, we need to set our sights higher than an analysis of the design of the latest high-end smartphone. We believe a very good place to start is with a review of the relatively short, yet thrilling, history of mobile handsets, providing an opportunity to understand the technological and market issues which have driven this phenomenal development.

1.1 Development of the first mobile handset

1.1.1 A famous telephone call

On April 3, 1973, Marty, a researcher at the US company Motorola, made a phone call from a Manhattan sidewalk to his colleague Joel Engel at the US telephone carrier AT&T.

The purpose of Marty's call that particular Spring day was to inform Joel, that he, Marty, was calling him from the world's first ever portable cellular telephone, beating AT&T in the technology race to develop a viable commercial portable cellular telephone. This first portable cellular phone was unlike anything we know today – consisting of about a kilogram of plastic and electronics, shaped something like a shoe, using analog radio technology, without any form of screen or menu buttons, and yet able to make and receive telephone calls "without wires" and when on the move. Marty, or, to give him his full name,

Martin Cooper, is now revered by many as the father of the mobile phone.

Ten years and five iterations of the handset design later, the US Federal Communications Commission (FCC) approved for use the world's first commercial portable cell phone on September 21, 1983 in support of the launch of the first commercial cellular network by AT&T in 1983. Weighing in at approximately 800 g, known as the DynaTAC 8000X, and with a ticket price of around $3500, the product was a true technological marvel. Within that first ten-year period from Martin Cooper's Manhattan phone call, the technology had already been miniaturized to approximately half the weight of the original working prototype. The vision which drove Martin Cooper, and a whole host of others, was the radical idea that people would much prefer the freedom to be contacted wherever they happened to be, and to be able to call other people without first needing to know where they were located. It is a reflection of how ubiquitous the mobile phone has become that today this statement seems almost "obvious," yet just 40 years ago this was a remarkable vision of the future of communications. Marty is the first to confess that significant teamwork by many thousands of people has been required to fulfill that vision, and that the vision is constantly being renewed as technological improvements make new capabilities available year by year. In the remainder of this chapter, we take a brief tour through the history of the development of the mobile phone, looking at some of the industry's key moments and reminding ourselves of how rapidly the technology has developed, how this has driven key design issues, and how ubiquitous mobile communications has become.

1.1.2 Early developments in mobile telephony

Before that historic phone call, the early development of mobile phones was characterized by a number of different, somewhat disconnected, inventions and experiments. For example, one Lars Magnus Ericsson (the founder of Ericsson, who retired in 1901 to go into farming) installed a fixed telephone into his "horseless carriage" (car) in 1910. As Lars and his wife Hilda traveled across the country, he would stop his car at various places and Hilda would connect two wires from the phone to overhead telephone lines using two long sticks.

Developments in two-way radio communication in the first half of the twentieth century were used in applications such as shore-to-ship communication, for example in 1926 for first class passengers on trains between Berlin and Hamburg, and for military communications during the Second World War. Following the Second World War, military and civilian patrol cars started to use two-way radio communications. In all these early examples, a mobile phone had to stay within a particular area serviced by a specific base station throughout the duration of the telephone call. This meant that mobility was severely limited, as there was no mechanism to transfer the call from one base station to another without first disconnecting the call, and there was no concept of re-using frequencies between different users. Solving these two problems of hand-off or handover between base-station cells, and the ability to re-use frequencies between phone users, were the key innovations which laid the groundwork for the first-generation mobile phones.

Radio telephone services commenced in the USA in 1946, for use primarily in vehicles. These so-called "car phones" were essentially two-way radios which allowed connection via a base station to the landline telephone system. Both the phones and the base stations transmitted at maximum power to ensure the largest possible area of coverage for making and receiving phone calls. The base station allocated two frequencies for the duration of each call – one for receiving and one for transmitting. Although successful, there were obvious limitations with this approach. Because of the high transmit power used, it was not possible to re-use the same frequency between adjacent base stations due to the interference caused. This, coupled with the fact that two frequencies were being used up for each call, meant that the maximum capacity of a base station was reached in a very short time. Nonetheless, the obvious value of being able to make phone calls from vehicles encouraged companies to invest in research and development to work through how to design better solutions which could handle much greater capacity.

1.1.3 The cellular innovation

In 1947, engineers at Bell Laboratories in the USA had proposed a hexagonal "cell" structure, with base stations transmitting (and receiving) in three directions at the "node" points where three adjacent hexagon shapes coincide.

The essence of the idea was that geographical areas would be divided into small adjacent cells, each operating at a lower power than a single large transmitter covering a much larger geography. Each of the cells would support a number of frequencies, not in immediate use by neighboring cells, which would allow many more two-way radios to be used at the same time. This was a radical idea well ahead of its time, as no technology existed to realize this idea practically for a further 15–20 years. In the 1960s, again out of Bell Labs, the transistor was invented and commercialized, with one of its many applications being to build electronics to realize portable communications with a small base-station cell.

A key advantage of the cellular design approach was to reduce the maximum distance a radio signal had to be transmitted, hence significantly reducing the transmitter power required as well as the amount of interference between adjacent transmitter stations. The ability to use lower-power transmitters also created the opportunity to create portable radio products with their own transportable power supply. Reducing the amount of radio interference also made it possible to utilize the same frequencies across multiple cells that were in the same region, and thus increase the capacity – meaning that more people could use the service at the same time in a particular region.

In 1968, the FCC first proposed to allocate frequencies in the 800–900 MHz range specifically for the deployment of new technologies, in order to solve the key issue of limited call capacity of the existing two-way radio telephone systems.

Research continued, and in 1970 Amos Joel, an engineer at Bell Labs, invented a system of "call handoff" which would allow for a mobile phone to be able to move between base-station cells whilst maintaining a continuous call connection, even if different frequencies were utilized by each of the cells.

1.1.4 Early commercialization in Scandinavia

It is fair to say that there are many "fathers" of the mobile phone distributed around the world. In Europe, Swedish electrical engineer Östen Mäkital's work in the 1960s led directly to the development of the first-generation NMT system in the Nordic countries, and some would claim Mäkital as the true father of the cellular phone.

In Europe, Ericsson's very first mobile phone was designed in 1956. It weighed 40 kg and was about the size of a suitcase. When mounted in a car, it cost almost as much as the car! Because the entire network for which it was designed could not serve more than about 100 subscribers, it was of limited commercial success.

In Sweden, in 1960, an early car phone system known as Mobile Telephone System A (MTA) was launched, with the car phones being provided by Marconi. The system provided incoming and outgoing call capability. Outgoing calls were made via a rotary dial. Incoming calls relied upon a human operator determining which cell the car phone was currently within and then routing the call manually to the appropriate base station. By the mid 1960s, MTA was replaced by MTB, and because of the introduction of transistors, it was possible to design smaller phones that required less power and were cheaper to manufacture. MTB was launched in 1965; however, it soon ran out of capacity, once it was serving 660 customers.

In Finland, in 1971, the ARP (Autoradiopuhelin, "car radio phone") system was launched. It was based on a cellular arrangement of base stations, although there was no handover between cells – leaving the coverage of a cell would cause the call to drop. The first ARP mobile terminals were large and could only be fitted in a car boot, with a handset near the driver's seat. In the 1990s, handhelds were introduced for ARP, but they never became popular as true cellular systems such as NMT became more ubiquitous with more affordable mobile phones.

1.1.5 The first commercial cellular launches

By the early 1970s, both AT&T and Motorola announced plans which envisaged the development and deployment of high-capacity mobile telephony networks based on the cell-based structure. AT&T submitted a proposal in 1971 to the FCC for the commercialization of a nationwide cellular mobile phone system. By February of 1973, Motorola had produced a working DynaTAC (Dynamic Adaptive Total Area Coverage) portable phone prototype, which they presented to the FCC along with design details for a complete cellular network system. The FCC agreed to hold new hearings on allocating spectrum for a future cellular service. However, the process of decision making at the FCC was very slow and bureaucratic. Agreement to a trial with AT&T was reached in 1975, though approval was not granted until 1977. Further progress was stunted by the larger debate about the future of the monopoly which was AT&T.

Records suggest that the world's first commercial cellular telephone system was actually launched in Bahrain in May 1978, operating with just two cells on 20 channels in the 400 MHz band and with a capacity of 250 subscribers. Telephones were supplied by Matsushita (Panasonic), and network equipment was supplied by Cable and Wireless.

In Tokyo, an 88-cell system was launched in 1979, using Matsushita and NEC equipment. The first North American launch was in Mexico in 1981, though this only operated on a single cell.

Nordic Mobile Telephone (NMT) System – a cellular system developed by Nokia, Ericsson and others, was launched in 1981 on a small scale (20 cells) in Saudi Arabia, followed by larger Scandinavian-wide deployment in Denmark, Finland, Norway and Sweden. Because compatible systems were deployed in each of these neighboring Nordic countries, NMT was the world's first network to provide an international roaming capability. The first NMT car phones were introduced in 1982 by Nokia Corporation. Several other countries also launched cellular networks in the early 1980s, including Mexico in 1981 and the first Vodafone network in the UK in 1985 using a variant system known as TACS (Total Access Communications System), followed by a number of other European countries – each with its own incompatible variant system. Note though that all of these early cellular systems operated using car phones – the story moves back to the USA for the launch of the world's first commercial handheld mobile phone.

After many years of deliberations, agreement was finally reached on the future of AT&T, leading to its breakup into a number of smaller regional operating companies in 1982. With this hurdle cleared, the regional Bell operating company Ameritech began operating the first commercial cellular service in the USA on October 12, 1983, in Chicago, with other launches following rapidly. The system was the AMPS (Advanced Mobile Phone Service) analog cellular system.

For Motorola, it had been a long wait. On September 21, 1983, Motorola was granted approval by the FCC to sell the DynaTAC 8000X phone – the world's first commercial portable mobile phone, after more than ten years of research and development, and perhaps more than a 100 million US dollars of investment. Motorola had recognized the potential to utilize developments in electronics and computing years before in order to create a much more portable wireless phone. Martin Cooper, who we met at the start of this chapter, had worked closely with Motorola's industrial design director, Rudy Krolopp, and his team, in order to conceive a design for the physical shape of the phone. A number of concepts were created, and a clear winner emerged, which, as Krolopp later recalled, "We called it a shoe phone, because it sort of looked a little bit like a boot."

Motorola DynaTAC 8000X

Voted one of the ten ugliest technical products ever by *PC World* magazine in 2007, the Motorola DynaTAC 8000X was the world's first ever mobile phone, and was a truly technological and product marvel when it was first launched in 1983, due to the utility value of being able to make telephone calls whilst on the move and out and about.

In 1984, the Korea Mobile Telecommunications Company was formed, and an AMPS cellular service was launched in the same year. Manufacturing of phones soon followed, with a joint venture in South Korea between Nokia and Tandy to form the Tandy Mobira Corporation. Tandy had a network of electronics stores across the USA, and Nokia wished to enter the US market. This was South Korea's first entry into manufacturing handsets, the first baby step in the direction of what would become a major manufacturing base for mobile phones, with the later emergence of globally significant players such as Samsung and LG.

Back in Scandinavia, in 1987 Nokia launched its first NMT handheld mobile phone, the Nokia Mobira Cityman 900. This phone gained the nickname of "Gorba" when the then Soviet leader Mikhail Gorbachev did a photo shoot making a call from Helsinki in Finland back to his Communications Minister in Moscow. The Mobira Cityman 900 weighed 800 g and had a price tag of 24 000 Finnish marks.

1.2 Generations of mobile communication capability

Before we move forward into the various generational advances from these first cellular network systems, it is worth our while to understand the key differences between the different generations or "Gs" in mobile communications. What a lot of meaning is carried in the implication of a move to "the next generation"! It was with the advent of the first digital mobile networks that the term "second generation," or 2G, was coined, and only post-rationally was the term "first generation," or 1G, applied to the preceding solutions. So it was that 2G provided a "clean slate" on which to design a new generation of cellular network systems based on advances in electronics which allowed much signal processing to be carried out by digital electronics.

The term "3G" was initially used to advent the move to systems which could support higher-speed data, and arguably to encourage national regulators to free up additional spectrum for use by future mobile communication services. However, in the meantime, a number of important incremental improvements were made to the 2G systems to support the move from traditional circuit-switched data to packet-switched data. With the birth of the World Wide Web in the early 1990s, mobile packet-data solutions became essential to provide an efficient method of carrying Internet traffic, which is itself packet-based. Having already allocated the term "3G," the industry resorted to labeling these important improvements as "2.5G" and "2.75G."

In the early 2000s the term "3G" was used to create perception in the mind of the consumer that they would be able to browse the Internet from their mobile phone using a high-speed data connection, much as they would from their desktop PC. Although this was technically possible, the user experience was pretty poor, and much work was still required to understand how to deliver Internet-based services and content to a mobile successfully. More of this later. After the over-hyping of the term 3G, there was caution in any use of the term 4G for a number of years, hence the use of terms such as 3.5G, 3.75G and even 3.9G! In 2010/11, US operators began to talk up the availability of so-called "4G" services to provide higher data rates and better capacity, as users migrated from feature phones to smartphones in large numbers. The major innovations brought by each generation may be summarized as in Table 1.1.

1.3 The digital revolution – 2G

With mobile phones becoming a proven commercial success by the early 1980s, plans began to be laid for successor systems. The number of simultaneous mobile phone users on the same cell – otherwise known as cell capacity – quickly emerged as a key issue as network traffic continued to grow. A successor system had to offer a significant improvement in capacity to continue to allow growth in the industry. Advances in the electronics industry, resulting in increasing computational capability and miniaturization without a corresponding increase in price, created opportunities to specify and create a significantly more advanced cellular system based on digital signal processing.

Advantages of digital over analog transmission are numerous. With rapid advances in electronics, greater levels of functionality became possible at a reducing cost. It became possible to integrate more functions onto an integrated circuit at a reduced cost and size. The move to digital provided many additional advantages, such as:

- the ability to use the available spectrum much more efficiently using modern information coding techniques;
- the ability to encrypt information, for example making voice calls secure from third-party interception;
- the ability to switch (and therefore share) frequencies more rapidly;

Table 1.1. *Generations of mobile communication standards*

Generation	Key innovations	Examples
0G	Wireless telephony capability between a mobile (moving) handset and a base station	Military use in WWII early commercial car phone systems
1G	Use of multiple small overlapping cells, resulting in better frequency re-use between cells and a reduction in the maximum required transmit power Ability to hand-off/handover calls onto different frequencies on different cells without dropping the call	NTT (Japan) NMT (Nordic countries) AMPS (USA) TACS (UK)
2G	Move from analog to digital radio transmission technologies Secure (encrypted) voice traffic Short messaging service (SMS) Circuit-switched data services (full duplex data and facsimile) International roaming (GSM)	GSM (Europe-wide and then increasingly globally) CDMAOne (North America, South America, South Korea, Japan, China) PHS (Japan)
2.5G	Packet-switched data services, more suited to the transmission of IP (Internet) based packet data	GPRS CDMA2000 1xRTT

2.75G	Faster packet-switched data services through improved modulation techniques	EDGE CDMA2000 1xRTT
3G	New CDMA-based modulation schemes to increase further data throughput rates	WCDMA CDMA2000 1xEV-DO
3.5G	Enhanced CDMA-based modulation schemes to increase further data throughput rates	HSDPA, HSUPA (HSPA)
3.9G	OFDM-based modulation schemes to increase further data throughput rates; often referred to by operators as "4G" or "pre-4G", though falling short of the recent ITU re-definition of 4G	LTE WiMax
4G	Changes to core network to support all-IP data transfer (even for voice calls) ITU October 2010 definition – speeds of 100 Mbps downlink with high mobility and 1 Gbps with limited mobility	LTE-Advanced WirelessMAN-Advanced
5G	Yet to be defined	

- the ability to introduce fast phone-to-network signaling in order to support advanced system features such as fast handover, authentication and encryption, international roaming, call forwarding and text messaging.

Other, economic and political, issues also drove the need for successor solutions. With a number of different analog systems launched, driven by various national agendas, a key limitation was the inability to use a phone which worked in one country when visiting another country. This was much less of an issue in the USA, with its large homeland geography, or in Japan with its distinctive culture. However, it was a critical issue in Europe, where, with the development of the European Economic Community (now the European Union) and with a number of adjacent countries with significant trade between them, there was a need and a desire to facilitate the free movement of goods and services – and people – between member countries.

From the supplier perspective, creating different phones for different network standards made the volume of product which could be produced for any one standard limited. Having a common standard would enable significant economies of scale by being able to create common products which could service multiple markets. Within the context of Europe, it started to make sense to conceive of a common standard that would allow a mobile phone to work in any country in Europe and would create a much larger market for these phones, in turn enabling costs to be driven down, thus creating an opportunity for many more people to be able to communicate seamlessly across borders. The vision of the European Union to enable the free movement of goods and services amongst member states provided important context for the idea of a ubiquitous mobile communications system for European citizens.

1.3.1 A common European standard

In 1978, European member states agreed to reserve spectrum in the 900 MHz band for future mobile communications. This agreement was reached through CEPT (European Conference of Postal and Telecommunications Administrations), a European body made up of the then state-owned national post and telecommunication organizations (PTTs). In 1982, CEPT launched a new standardization working group called the Groupe Spéciale Mobile (GSM), transferred subsequently in 1989 to the new European standards group ETSI (European Telecommunications and Standards Institute). ETSI had the advantage of accommodating equipment makers and technology suppliers in addition to telecommunication operators, thus allowing the companies most associated with innovation in mobile telephony to participate. By 1991, the GSM standards – known as recommendations – amounted to 130 documents with over 5000 pages of content.

A key decision for the GSM working group was the selection of the method for accessing the radio spectrum for the transmission of signals – known as the method of channel access. The existing analog networks utilized frequency-based channel access (different conversations on different channels – FDMA). GSM favored a time-based channel access known as time division multiple access (TDMA), which allowed the multiplexing of a number of channels onto the same frequency by providing for each channel to "have a turn" at transmitting or receiving before giving the next channel "its turn." In GSM, each frequency is divided into eight repeating time slots, supporting eight channels. For TDMA to work, it is crucial to have a very accurate "time keeper," with all mobile handsets communicating with their base station to stay in synchronization with each other. This challenge could only be overcome with the move to digital technology, as data could be sent as a discrete number of bits which would fit within a known period of time, and data coding techniques could be employed to ensure that handsets stayed in synchronization with their base station by "listening" to a timing signal and staying in step with it over time.

1.3.2 Digital developments in the USA

As subscriber numbers continued to grow dramatically in the USA, the search began in the late 1980s for a replacement system for AMPS that could deliver higher capacity. In 1988, the CTIA (Cellular Telecommunication Industry Association) published a set of requirements which sought a ten-fold improvement in capacity through the use of digital technology. In 1989, the TIA (Telecommunication Industry Association) selected a TDMA-based approach to provide a new digital cellular system with the ability to provide an approximate tripling of capacity. Although the CTIA went along with the TIA recommendation, the desire for a ten-fold improvement in capacity was not inherently possible, although the CTIA hoped that future improvements could deliver these gains. In 1990, IS-54, or D-AMPS, was formalized; this was based on TDMA, and was able to work with existing analog AMPS systems. With the move from analog to digital, capacity was increased by sampling data at a very fast rate and multiplexing conversations onto the same physical channel.

Meanwhile, Qualcomm, a wireless technology company largely working in the field of satellite communications, was exploring a new approach to channel access for cellular phone systems by building on its expertise in an access method well utilized in military and satellite communications. This method is known as CDMA (code division multiple access) and develops the idea of *spread spectrum* – a method of spreading out the information from a signal (determined by a code value) that is to be transmitted across many different frequencies over time (rather than across a single frequency). If the receiver knows which code to use, then it can determine at which frequency, and at what time, to receive and decode the transmitted information, allowing it to reconstruct the original signal.

Because the signal is spread out over many different frequencies, the probability of interference is much reduced (it is more likely that some frequencies will suffer interference than all available frequencies), and the signal being transmitted is inherently very secure (because you need to know the code sequences being used in order to listen in to the signal). The more rapidly that the signal can be spread across the available frequencies, the more information is transmitted, and thus the greater the capacity (number of simultaneous users) that can be achieved.

CDMA had already successfully been deployed in the fields of satellite and military communications, although no one to date had found a practical way of applying CDMA to cellular systems. Simulations by Qualcomm showed that CDMA could provide many more phone calls for the same spectrum as other systems. However, there were a number of known issues with the application of CDMA to cellular solutions that were generally felt in the industry to be difficult to surmount.

These shortcomings included the well known *near–far field effect.* Simply put, this means that the further a mobile phone is away from a base station, the more transmit power it needs in order to be heard by the base station (and vice versa). If a mobile phone is near to a base station, but uses the same amount of power to transmit as if it were further away, it will effectively "shout," causing interference on other adjacent frequencies. In a CDMA system, such "shouting" would cause loss of signals on many other frequencies – because any one signal is "smeared out" across many frequencies over time. An effective system would need to be able to modify the transmit power – very rapidly – depending on how near or far the mobile phone was from the base station. Because in CDMA the signal to be transmitted is rapidly being "chopped up" into smaller amounts of signal which are transmitted across a broad spread of frequencies, there were huge technical challenges in being able to adjust the transmit power rapidly enough. A further problem was in handover between cells. In existing AMPS systems, the approach was a "break and then make" – meaning that, in handover, the existing call was ended briefly and a new call quickly set up on the new frequency on the new cell. With CDMA, a break in communication, however briefly, could cause significant problems in re-establishing communications on the new channel, due to the fine timing required to "code" the data and spread it across many different frequencies.

During 1989, Qualcomm was able to approach each of these shortcomings in the use of CDMA for a cellular system, and, one by one, to develop solutions which overcame these perceived shortcomings, whilst continuing to provide the benefits of a significant increase in call capacity over the existing AMPS system. A significant part of the power control problem was solved by using an existing circuit on a mobile phone known as the automatic gain control (AGC), which adjusted the voltage level of the incoming signal up or down to provide a consistent voltage level for other circuits involved in decoding the signal. The incoming voltage level is related to the strength of signal being received, which in turn is related to the distance between the base station and the handset. Using this information, the handset could adjust its transmit power up or down very rapidly to reflect its distance from the base station. Coupled with measurements at the base station of mobile phone signals and instructions to individual mobile phones to "turn their signal up a bit or down a bit," Qualcomm effectively solved the power control issue.

The handover issue was solved using a two-part solution. The first part was to ensure that the timing signals used by each base station were synchronized using a single master clock (provided by GPS). The second part was to ensure that, when handing off from one base station to another, both base stations transmitted exactly the same signal to the mobile during the handover process. This meant that a mobile could switch from one base station to the other without any loss of signal. This handover approach was dubbed a "soft handover" rather than the more "brutal" approach of existing AMPS networks to break the call on one cell and then (quickly) re-establish the call on the new cell. An additional advantage therefore of the soft handover approach was the lack of an annoying "click" during the handover process.

Having proven the suitability of CDMA for cellular systems, Qualcomm was able to work with a number of network operators, first in the USA and then internationally, to deploy CDMA-based systems as an alternative 2G solution to the "conventional wisdom" of TDMA-based systems deployed by GSM in Europe and D-AMPS in the USA.

1.3.3 Launch of the first 2G networks

The first GSM network was launched by Radiolinja in Finland in 1991. The first CDMA networks were trialed in 1993, with the first commercial launch by Hutchinson in Hong Kong in 1995. In both cases, these new digital systems made use of existing spectrum allocations, displacing channels previously used for 1G systems. With rapid growth in 2G systems, the earlier 1G systems were closed down to make more capacity available for the newer 2G systems, and users transitioned over to the new systems.

Continued advances in silicon integration had dramatic impacts on handset form-factor – size and weight, and improvements in battery life. Further battery life gains also occurred as a result of the smaller cell sizes introduced to accommodate a growing population of users, particularly in urban areas.

Nokia 2110

The Nokia 2110 was first launched in 1993, and quickly became a best seller. The outstanding feature of this phone was its intuitive dual soft-key user interface, which made the phone's advanced features much easier to access and use. In time, this user interface would be rolled out over most of Nokia's product range, and became a significant differentiator for the company's products. The ease of use of the product gained Nokia much respect and a following amongst consumers, with many users of the 2110 staying loyal to future Nokia products because they knew how to operate them. The 2110 was an early example, in an industry driven by technology, of the business value of good user experience design.

Second generation systems also introduced a new mode of communication – short message service (SMS), or, more colloquially, text messaging or "texting." Initially specified for GSM, SMS was eventually rolled out on all digital networks. Originally, SMS was conceived as a solution looking for a problem. As the GSM standard developed, there became an opportunity to utilize spare capacity on the signaling channels to send a small data message between the handset and the network and vice versa. At this time, no one was really clear on how such a capability would be utilized. It has also been reported that a pruning exercise was undertaken, when the specifications were being finalized, with SMS being one of the candidates considered for removal from the specification. Luckily, SMS was not pruned – although in the early versions of the standard, SMS was only mandatory for messages sent to the handset – there was no need to be able to send messages back. This underlies the case that no one in the early 1980s could possibly conceive of the level of interest which would be unleashed once consumers worked out what they could do with SMS.

The early launches of networks often utilized (as today) SMS as a notification method to the handset for the arrival of new voicemail. Many of the networks did not charge at all for SMS in the early days, as usage was low, and the business case for investment in message switching centers was unclear. Uptake of SMS was very slow to start with – even with two-way messaging – because it was not possible at first to send an SMS between network operators. Over time, the network operators collected evidence that indicated that people were beginning to use SMS to communicate with each other, and the case grew to connect the short message service centers run by the individual operators, so that it became possible to send a text message to any person with a mobile phone. Once these inter-connection agreements were in place, and after the launch of pre-paid contracts, SMS went through a very rapid growth phase. This growth was particularly notable amongst the younger generation, who found SMS a cheap and highly convenient way of staying in touch with their friends; they created their own abbreviations to cram more information into a message, which had the added benefit of bewildering the older generation!

The ability to access media content on mobile phones was also introduced by 2G. In 1998, the first downloadable content sold to mobile phones was the ring tone, launched by Finland's Radiolinja (now Elisa). Advertising on the mobile phone first appeared in Finland when a free daily SMS news headline service, sponsored by advertising, was launched in 2000.

Mobile payments were trialed in 1998 in Finland and Sweden, where a mobile phone could be used to pay for Coca Cola from a vending machine and car parking. Commercial launches followed in 1999 in Norway. The first commercial payment system to mimic banks and credit cards was launched in the Philippines simultaneously in 1999 by mobile operators Globe and Smart.

The first full Internet service on mobile phones was introduced by NTT DoCoMo in Japan in 1999.

1.3.4 How the market developed in the 1990s

For the first five years after the launch of GSM and CDMA networks, the focus for innovation in handset design was around reducing cost, size and form-factor, and improving voice quality and battery life. In this respect, product innovation was significantly focused on making the core technology do a better job, and, riding on the back of Moore's Law, which allowed greater levels of silicon integration to deliver lower power, smaller size and lower cost. At this time, the major handset manufacturers such as Motorola, Ericsson, Nokia and Siemens (the so-called "MENS club"), were fully integrated businesses, investing significant R&D in developing their own core technology such as chipsets and protocol software, as well as developing ranges of handset models utilizing the resultant in-house technology. Soon, however, with the early growth of 2G systems, market conditions were ripe for a significant second tier of manufacturers to enter the market and seek to gain market share. Companies in this category included many European companies focused on GSM, such as Alcatel, Bosch, Cetelco, Dancall, Philips, Sagem and Telital. In Japan, established consumer electronics companies created products for the domestic market, including Fujitsu, NEC, Panasonic, Sharp, Sony and Toshiba. In the emerging tiger economy of South Korea, new market entrants invested in mobile included Daewoo, Lucky Goldstar (LG) and Samsung. These second-tier manufacturers, without the resources to create their own chipsets and software stacks, worked with merchant silicon providers such as Analog Devices, Lucent, Qualcomm and TI, along with then independent providers of protocol stack software such as Condat and Optimay in Germany and TTPCom in the UK. Companies such as Motorola, Philips and Siemens designed chipsets internally, though later they spun these chip divisions out as separate businesses to supply silicon to other handset manufacturers – Freescale (2004), Philips Semiconductors (2006) and Infineon Technologies (1999), respectively. The strategy behind supplying other handset companies, who potentially competed with the parent company's own handsets, was to drive scale, grow market share, and ultimately to create a more competitive price point for the handset division.

Nokia 3110

As the mobile phone market continued to grow rapidly and prices fell, owning a mobile phone became a viable consumer proposition, and so attention turned towards making phones much easier to use. Through the 1990s, Nokia stood out as the one organization which understood the importance and value of a very good user interface. A significant innovation occurred with the launch of the 3110 in 1997, which heralded the Navi-key single soft-key user interface. Along with a Clear key and an Up and Down key, the user interface of this phone was minimalist in design. The single soft key contained the most likely "forward" action of the phone. So, for example, after entering the digits of a telephone number, the soft key label would change to "Call." Pressing the soft key would initiate call dialing, and then the soft key legend would change to "Cancel" whilst connecting or "End" once connected. With this approach, it was possible for the novice mobile phone user to make use of all of the main features of the handset quickly and easily.

In 1993, just two years after market launch of the first GSM networks, the first one million user milestone was reached. Three years later, in 1996, the total number of users exceeded ten million. Growing market volumes created improving economies of scale, which began to repay the large investments in R&D, whilst reducing production costs, by keeping factories fully utilized. With each iteration in chipset design bringing numerous benefits in cost, size and power consumption, time to market began to emerge as a key market differentiator. If a manufacturer was late to market with their product, they could easily find themselves launching with a product that was already uncompetitive when compared with other devices on the market. In addition to the significant R&D costs for chipset design, software rapidly emerged as a key determinant of time to market and, to some degree, product cost. The software "footprint" required to create a 2G handset was at least ten times more for the early GSM phones compared with the previous analog phones. However, the complexity of the software increased dramatically with the move to 2G, resulting from the need to undertake complex real-time signal processing as well as advanced signaling protocols to support the complex interactions between the phone and the network to manage information transfer, mobility and service provision.

Motorola StarTAC

Launched in 1996, the StarTAC from Motorola was perhaps a first at establishing the principle that the overall design of the handset (from an aesthetic perspective) matters at least as much as the functionality of the product, and it quickly became the aspirational product to own, with many appearances in films of the time. The phone was a marvel of miniaturization, being both the smallest (89 mm × 51 mm × 19 mm) and lightest (88 g) phone in existence at the time. The phone was also wearable, as it could be easily clipped onto a belt. The "clamshell" design was the first of its kind, at least in the USA, and it was one of the first phones to introduce a vibrating alert, providing an unobtrusive silent ringtone. A unique feature was the ability to have two batteries fitted at the same time. The StarTAC was launched initially on the analog AMPS network in the USA, with later variants supporting TDMA, GSM and CDMA. In 2005, PC World voted the StarTAC number 6 in its list of the top 50 gadgets of the last 50 years.

1.3.5 That other disruptive technology – the World Wide Web

The business case for 2G was fundamentally built around driving subscriber numbers through improved network capacity and cheaper handsets. The "killer application" was voice, the "killer use case" was being mobile – typically away from the office. Although 2G systems such as GSM did support the capability for data services, the maximum data rate supported was 9600 bits/second, and there was no "killer use case" which needed data. At this time, the only successful mobile data systems were custom-built data networks providing service to businesses in the transportation sector, such as the trucking industry or regional taxi companies.

Outside the mobile industry, the disruptive innovation of the World Wide Web was launched on the world in 1991. Tim Berners-Lee, a researcher at the particle accelerator center at CERN in Switzerland, created a set of software protocols which provided a simple way for documents and other files on one computer to be accessed from a different computer. This was achieved by quoting a reference to that document, known as a Universal Resource Locator, or URL. These URLs were a type of hyperlink, that is a way of connecting or linking between files using an intermediate name, rather than needing to know exactly where the file was stored. Berners-Lee also created a markup language (HTML) which was readable by computers, and just about readable by humans, which described how to lay out (or render) information such as text and images on a screen. In addition, and critically, HTML allowed hypertext links (URLs) to be inserted, providing jumping off points to reference other information stored on other pages, quite possibly on other computers. By separating out where the information was stored, how to describe the layout of the information, and where to render (display) it, Berners-Lee created a powerful method that enabled people to publish information on networked computers that could be read by other people from any other computer connected to the network. Tim nicknamed the capability he had created "the World Wide Web" – due to the multitude of links which could now be connected between different information across the whole network of computers. The World Wide Web was announced to the world during the summer of 1991. At this time there were fewer than ten million users globally of the Internet. Within ten years this had grown to over 100 million users connected to the Internet, and by 2011 there were an estimated two billion Internet users globally. Although the Internet existed for many decades before the World Wide Web, it was the web which made the Internet instantly usable by millions, and now billions, of people.

Sharp J-SH04

Launched in Japan in 2000, The Sharp J-SH04 was the world's first phone with an integrated digital camera. Its successor, the J-SH05, was the first mobile phone with a 65 536 color TFT LCD. These phones marked the beginning of a dramatic change in the use of phones – beyond "talk and text" devices towards the rich multimedia personal computing devices which we are so familiar with today. Within two years of the launch of the J-SH04, more than half of the

Japanese subscriber base had a camera phone, and by 2003 more camera phones were being sold worldwide than stand-alone digital cameras. The world started to look increasingly to Japan for future device innovation, as Japanese consumers led the way in driving demand for ever more functionality.

Second generation mobile systems had been conceived in the mid 1980s and launched in the early 1990s – about the same time as the web was born. There now existed the beginnings of a possible killer use case for mobile data – the ability for people to access these same websites when on the move, when out of the office, to access a growing set of information that was being put "online" – on the web. The difficulty was, however, that 2G systems had been designed on the telephone industry paradigm of a circuit – that is, a continuous connection between the phone user at one end of the conversation and another phone user at the other end of the conversation. This was the model used for the wired telephone system, and this model was extended into mobile. Circuits were *switched* between physical bearers, such as landlines and wireless, which allowed the re-use of resources when no longer required. However, resources were allocated, for the duration of a phone call, whether or not anyone was speaking. For voice, this was generally not too great a problem, as usually either user (or sometimes both!) would be speaking. However, when we consider data traffic, we find that the nature of communication is rather different, with data tending to be much more "bursty" – meaning that there are short periods of a lot of communication, followed by longer periods of no communication. If we consider the case of viewing a webpage, there is a small amount of data going in one direction which communicates "get me this page," followed by typically a larger amount of data transfer in the other direction, where the data representing the webpage is sent to the recipient. There is then often a period of no data transfer, where the human at the receiving end is reading the webpage. Hence, there is clearly an opportunity to re-use the available resources (such as the air interface) for other purposes during those times when data is not being sent.

In the early 1990s, a new set of mobile standards was drafted to overlay a packet-based transmission system onto the existing circuit-switched mobile telephony standard, which by this time was witnessing phenomenal growth and success. The mobile industry roadmap envisaged the next major innovation in mobile telephony to be a third generation of cellular telephony, supporting much higher data rates, sufficient to allow advanced services such as video telephony. This was still fundamentally a person-to-person, circuit-switched view of the future. In response to the emergent opportunity to provide access to Internet-based services from mobile devices, the so-called "2.5G" standards such as GPRS and CDMA2000 1xRTT were created. Taking the example of GPRS as a case study, two fundamental changes were made to accommodate packet-based mobile data. These were, firstly, overlaying changes to the air interface to support higher data rates, and secondly designing additional packet-oriented network components to sit alongside the existing circuit-switched networks, without requiring significant changes to existing network infrastructure, which was expensive and had taken many years to roll out.

As we have seen, GSM is a TDMA-based system which divides the time domain of a channel up into eight repeating timeslots, with one circuit-switched telephone call allocated to a particular slot. The specification of GPRS allows various combinations of consecutive slots to be used together to carry the same packet-switched traffic. A different number of slots could be used on the uplink (traffic from the handset to the network) and the downlink (traffic from the network to the handset), allowing asynchronous uplink and downlink data rates to be supported, reflecting the nature of typical web traffic.

Significant signaling changes were required between the handset and the network to control the establishment and management of data sessions, which are different in nature to circuit-switched connections, and an overlay of network elements was required to manage the routing of data packets both within the network and, more fundamentally, to interwork with the Internet as the new backbone for movement of data.

From a handset design perspective, the move to packet-based systems heralded numerous changes to all aspects of the system design.

At the chipset level, the main change was the need for faster processors and fast access to much more memory in order to handle the increased flow of data off the air interface, needing decoding, assembling into a data stream, and then further decoding and rendering of the media content being transported. Whilst this was good, as it drove demand for more advanced chipsets and allowed chip manufacturers to invest with confidence, it was to be several generations of chipset evolution before device capability made a reasonable match with consumers' expectations of what a mobile Internet experience should feel like.

At the software protocol level, a brand new protocol stack was required to coexist alongside the existing 2G protocol stack. Early 2.5G products were PC card data modems, which interfaced to an application client – typically a laptop computer – via a traditional wire line modem "AT interface" with extensions for cellular.

At the software application level, there were huge challenges in meeting the opportunity of the "Internet on your phone" with the realities – both technical and commercial – of what was currently achievable. It was all very well providing an Internet connection to a mobile device, but devices of the late 1990s and early 2000s were still very constrained in terms of processing capability, memory costs and display capabilities. Some pioneering products were created with memory and processor efficient implementations of a full HTML browser on a phone, but, although a technological marvel, they were ahead of their time in terms of the ability of all the other mobile handset technologies to provide a good user experience at an acceptable price point, as the industry was yet to respond wholeheartedly to the opportunity of mobile Internet devices. For example, at the time, chipset clock speeds were measured in tens of megahertz, memory in the hundreds of kilobytes, displays were monochrome, and networks were just beginning to migrate from circuit-switched data at 9.6 kbit/s to the newer packet-switched GPRS networks at perhaps 56 kbit/s.

RIM BlackBerry 5810/20

The RIM BlackBerry device first arrived in the USA in 1999, as a pager-like wireless messaging device, running on the Mobitex data-only network in the 900 MHz band. The product had a monochrome alphanumeric display and an unusually arranged Qwerty keypad, ergonomically designed to allow individual letter keys to be pressed, whilst containing the design into as small a form-factor as possible. The BlackBerry 5810 and 5820 devices, shipped in 2002, were the first RIM BlackBerry devices to work on cellular networks, in this case the GSM-based networks of North America and Europe, and provided the mobile operators with a significant benefit which their new GPRS data-enabled mobile networks could offer to corporate users. Back in 2002, these devices were positioned as wireless PDAs (personal digital assistants), as opposed to smartphones, due to their more data- and computer-centric uses. Indeed it was necessary in these early models to use a hands-free headset in order to use the device to make phone calls.

The mobile industry believed it had the answer through the creation of a new set of "stripped down" Internet protocols, which could deliver an acceptable user experience given the technological roadmap. A new set of wireless Internet specific protocols – known as WAP, or Wireless Application Protocol – were developed in parallel to changes to the core network. Each level of the Internet and web protocols was replaced by a wireless protocol equivalent which required less network bandwidth and less processor bandwidth than the "true" Internet protocols, and was more suitable for rendering on the small, typically black and white, screens of the day. An additional WAP gateway function was required in the network to convert between the Internet protocols and the mobile-specific WAP protocols. The first WAP phones were launched in 1999, amidst much industry acclaim that they were heralding in the ability to surf the new information highway from your mobile handset. Unfortunately, expectations were set far too high, service design was generally poor, and the revenue share model between operators and developers was heavily weighted in favor of the operators.

Nokia 7110

Launched in 1999, the Nokia 7110 was the first commercially available WAP phone, also notable for its distinctive curved shape (often known as "the banana phone"), which was an iteration on the earlier 8110, featured in the classic science fiction film *The Matrix*. The 7110 demonstrated the commercial viability of WAP with access to sites providing information such as stock prices, exchange rates, news, weather, sport, flight schedules and hotel bookings. However, this product only supported a theoretical maximum data rate of 14.4 kbit/s over a circuit-switched connection, as GPRS packet-switched capability had not yet been rolled out in the networks.

Separately, in Japan, DoCoMo, a network operator, took an alternative approach and created a stripped down version of HTML, known as compact HTML or cHTML. In February 1999, DoCoMo launched a mobile Internet service in Japan called i-Mode; i-Mode was to become phenomenally successful, achieving its first million users by August of the same year, with a rapid growth rate leading to 40 million users by October 2003. By contrast, WAP services struggled to take off due to a less integrated solution approach between operators, original equipment manufacturers (OEMs) and content providers, a large gulf between the expectations set around surfing the mobile Internet, and the reality of text and simple graphics on small black and white screens with slow data connections.

Sharp GX10

The Sharp GX10 was the first mass-market color screen camera phone available outside of Japan, and was created in close cooperation with the operator group Vodafone for the launch of their new mobile

Internet portal known as "Vodafone Live!" Vodafone's Live! service was of itself a more radical attempt by a network operator to package the "raw" technology of WAP and GPRS packet data into an easy to use consumer-oriented service experience, seeking to deliver on at least some of the promise of the "mobile Internet" experience heralded by many operators' marketing campaigns.

The GX10 was innovative and iconic in bringing a number of relatively new premium features to the European market – such as a high-quality screen, digital still camera, Java execution environment for applications and polyphonic ringtones. Further, the handset application software was highly customized and branded for Vodafone, to an extent which had not been achieved before outside of Japan. In the Japanese market, a very close relationship between handset suppliers and network operators had developed, which permitted the network operators to have a very strong say in the specification and branding of handsets.

It appears that i-Mode was successful for four key reasons. Firstly, the users of i-Mode were largely younger people whose only means of access to Internet-based services was via their phones. (Most Japanese young people did not have their own private space, with access to PC or TV.) Secondly, the i-Mode revenue model was innovative, with a revenue-share arrangement favoring developers over DoCoMo themselves. Thirdly, DoCoMo was able to control the handset specifications from its OEM suppliers very tightly, leading to a well integrated overall service experience. Finally, buoyed by initial success, DoCoMo was able to use i-Mode as a platform to roll out further innovations such as Java, music, video, location, etc., which, along with its other domestic operator competitors, led to Japan taking the global lead in mobile services and next-generation handset features for many years. DoCoMo sought to export its i-Mode success to other affiliated operators such as O2 in the UK and Telstra in Australia. However, these services were never anywhere near as successful as their Japanese counterparts, as the service experience was just not a match for competing high-speed fixed-line Internet access available in Western geographies. A lack of confidence in the market led to only a few handsets being available for any one network operator, as OEMs were only prepared to make limited investments.

Like i-Mode and cameras, the idea of putting a Java execution environment on a handset to allow applications to be downloaded was first pioneered in the advanced market of Japan around 2000. Within less than five years, Java became a mainstream requirement for most feature phones, and enabled for the first time the ability to develop third-party applications for mass-market handsets. The standards for Java on handsets evolved in an uneven way, and one of the big challenges was defining a baseline of reasonable functionality that would be supported on the majority of Java-enabled handsets. This led to fragmentation in the market, which slowed the success of the Java-based approach. An alternative approach was provided by Qualcomm in 2001, with the launch of BREW – an execution environment for handsets, with the back-end infrastructure to host applications in a store and a payments settlement system to ensure that all parties were paid. BREW is a white label solution, which was licensed, in time, to over 60 network operators globally. BREW was probably the first fully integrated application store solution available globally, an approach which was famously re-invented by Apple with the launch of the iPhone and the Apple app store in 2007.

Motorola Razr

Launched in 2004 as a very thin profile clamshell phone, the Motorola Razr was originally positioned as an exclusive fashion phone. Because of its iconic design, it quickly became very popular, and, with a reduction in price, was to go on to sell over 50 million units by 2006. Developed by a team in Motorola with the brief to think differently from the rest of the organization, it marked a discontinuity in the market away from technology-led products to design-led ones. Although intense attention was given to a unique physical design, the user interface software was largely re-used from existing products, and many users reported the product difficult to use for this reason.

1.4 High-speed data – 3G

The main focus of third-generation systems is on higher data rates in excess of 200 kbit/s. There are three main 3G standards. These are UMTS, standardized by 3GPP, CDMA2000, standardized by 3GPP2, and TD-CDMA, used in China. In April 2000, the UK government's auction of spectrum for new 3G licenses concluded, having raised £22.5 billion ($US34 billion, or 2.5% of UK GNP). In August of the same year, the German spectrum auction exceeded this by raising £28 billion ($US42 billion). There has been considerable argument about whether the telecoms companies paid too much for these licenses. Whatever one's views on this, it is certainly the case that operators had the imperative to secure 3G licenses in order not to be locked out of the next wave of technology innovation, and that it has taken a decade following these licenses for 3G to become mainstream. By 2011, approximately 50% of subscriptions in Europe are 3G, although globally penetration in 3G varies considerably from Japan, with over 90% penetration, to China and India, with only single-digit penetration.

Apple iPhone

Released in 2007, the Apple iPhone is well documented as creating a major discontinuity in the industry. Apple's design-led and user-focused approach to products and services was the first properly to deliver on the long-promised industry vision of advanced handheld products which consumers would use to access a wide range of online services and information, with a corresponding "always on" and "always connected" lifestyle. The irony for the network operators who had long led this vision, was that Apple's business model began a process of moving customer loyalty and mindshare from the operator as service provider to new players such as Apple, Google and Amazon, who had developed strong services through the somewhat separate and parallel Internet boom.

1.5 Mobile broadband – 4G

The two 4G standards are known as LTE and WiMax. Both are distinguished from previous 3G standards by two key advances. The first is an all IP (Internet Protocol) network from end to end. All traces of legacy circuit-switched capability have been removed, even for voice calls, meaning that voice is carried as a series of data packets at all points through the network, just as all other types of data are. The second key difference is that both LTE and WiMax use OFDMA (Orthogonal Frequency-Division Multiplexing Access) as their channel access technology, which permits much higher data rates to be supported.

The meaning of 4G, however, has had, and continues to have, a rather fluid meaning. The ITU in 2010 re-defined the target for 4G as speeds of 100 Mbit/s downlink with high mobility and 1 Gbit/s with limited mobility. Under these recent definitions, only the future evolution of LTE known as LTE-Advanced and the future evolution of WiMax known as WirelessMAN-Advanced now qualify as true 4G systems, even though both LTE and WiMax have been positioned as 4G solutions for many years. At exactly the same time, network operators in the USA have been rushing to launch higher data rate networks, all billed as 4G, but using a wide range of (now reclassified 3G) standards such as the T-Mobile USA HSPA+ network, Clearwire's WiMax network offering average speeds of 3 Mbit/s to 6 Mbit/s, and Verizon Wireless with its LTE network launched in early 2011 offering speeds between 5 Mbit/s and 12 Mbit/s.

1.6 Conclusion

The mobile handset has undergone rapid evolution in its short history. The pace of such change has been set largely by advances in semiconductor technology, which permit more and more capability to be embedded into silicon at lower and lower cost. As prices fall, new markets become available, and this drives volume and manufacturing efficiencies, further reducing cost and creating a virtuous circle. As the number of users on networks continues to grow, network capacity and information throughput remain the key design issues for new communication standards. With falling prices and increasing performance, new market segments for mobile handsets have continued to open up, leading to rapid innovation in product design, usability and product capabilities. Rapid convergence of telecommunications, digital media and the Internet economy, with the resulting emergence of a connected, digital lifestyle for billions of people around the world, means that the mobile handset and its derivatives remain the most dynamic and widespread consumer electronics products ever. As technical complexity continues to increase, a constant challenge for handset designers is simplicity and elegance in the overall product experience – these are the products which ultimately are remembered as iconic.

In this chapter we have seen how handset technology, the global marketplace and overall device experiences have changed rapidly over just a few decades. In the following chapters, we examine a wide range of topics which cover the key issues that contribute to the design of a modern mobile handset. As we undertake this exploration, we shall continue to maintain a perspective on how the mobile handset has developed over time, as we believe this allows us to bring into sharper focus key design issues which remain relatively constant within an industry of constant change.

1.7 Timeline of the mobile phone

We include the timeline from the first portable cellular telephone call in 1973 until the present time.

1973 Martin Cooper makes the first portable cellular telephone call

1979 First commercial cellular network launched in Japan by NTT

1982 NMT analog cellular system launched in Denmark and Finland. Roaming between networks in Sweden, Denmark and Norway introduced

Groupe Spéciale Mobile (GSM) is set up from an initiative by CEPT (European Conference of Postal and Telecommunications Administrations) to create a common European digital cellular communications standard

1983 AT&T launches the first US cellular network based on the AMPS standard

Motorola launch the world's first commercial handheld cellular mobile telephone – the DynaTAC

1985 Vodafone made the UK's first mobile call at a few minutes past midnight on January 1, 1985

1989 Groupe Spéciale Mobile (GSM) is formed within the European standards group ETSI, with a target launch date of 1991

Motorola MicroTAC is launched as the world's smallest and lightest mobile telephone

1990 GSM finalizes the first set of specifications for the new GSM system

1991 Vodafone launches the first GSM service, in the UK.

Widespread shortage of mobile phones until 1992

Reputedly the first SMS is sent on December 3 on the Vodafone UK network

1993 NTT DoCoMo launches 800 MHz digital cellular phone service

1997 On June 11, Philippe Kahn instantly shares the first pictures from the maternity ward where his daughter Sophie was born, with more than 2000 members of his family, friends and associates around the world. This is the first publicly known and shared picture

1998 In Finland and Sweden a mobile phone is used to pay for Coca Cola from a vending machine and for car parking, respectively

1999 NTT DoCoMo in Japan launches i-Mode, a service for accessing information on the Internet from a mobile device

The first WAP-enabled phone, the Nokia 7110, is launched

2000 3G spectrum auctions in UK and Germany

First commercial camera phone, Sharp J-SH04, launches complete with infrastructure for sending photos from an integrated CCD sensor

2001 First 3G test call made in England by Vodafone

First BREW application store launched by KTF in South Korea

2002 One billion mobile subscribers milestone is reached

Sony Ericsson launch T68i – the first MMS-capable mobile handset

2003 Network operator Three launches the first 3G service

2004 Motorola Razr launched with focus on iconic design features

2007 First-generation iPhone launched along with App Store

Android operating system launched for developers

Amazon Kindle, first wireless e-reader, launched

2008 First commercial Android device launched (HTC Dream)

2010 First iPad launched, establishing a new "tablet" class of wireless consumer device

2011 First commercial LTE network launched in North America

2012 Instagram, a service for sharing photographs captured on mobile devices, purchased by Facebook for $US1 billion

2 Design influences

In this chapter we examine the most significant factors that good design teams consider when creating a new mobile handset design. The consideration of these factors and the design decisions taken have a major impact on the nature and success of the mobile handsets which are launched into the market. Designing a handset could be thought of conceptually as similar to cooking with a master chef putting together a new recipe. Different ingredients (design handset influences) are fused together to create the overall flavor of the dish (mobile handset) depending on the style of cuisine the customer is looking for (e.g. smartphone user or ultra-low-cost handset user). Underpinning the success of a good meal is usually the style and personality of the chef who is willing to take risks, innovate and try out new ingredients and methods of cooking, and at times to challenge traditional methods of cooking. This is no different to mobile handset design, where the culture, leadership and ability to challenge conventional thinking can contribute to new ways of creating and experiencing a mobile handset. We will explore the key design influences on mobile handset design, based on a framework of key contributing factors. We believe that careful consideration of these influences leads a design team to a position where they can successfully answer a very important question: "Why *should* users use this particular mobile handset versus any alternative?" The strength of the answer to this question is a measure of the "engagement value" of the resultant design.

2.1 Core design influence – engagement value

When commencing a new handset design, the first priority is to decide how to embed deeply into the handset the experience the designer would like the user to have throughout their use of the handset. The experience begins at the first sight of the handset through advertising and promotion,

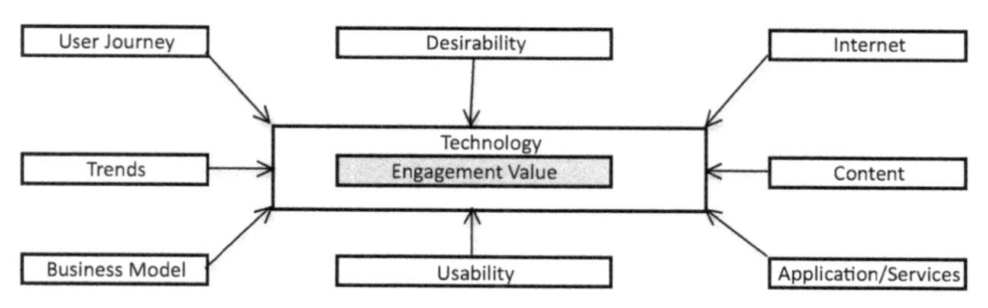

Figure 2.1. Factors influencing the design of a mobile handset.

and continues through the purchase experience, first use, then daily use, including downloading of applications and access to services. Finally the experience should be extended to the process of disposing of the handset, and hopefully then buying the next new handset from the same manufacturer. This process of designing for experiences also extends to other customer touch points, for example how the user experiences a technical problem with the handset and uses on-screen instructions or phones a call center for help. Each point of interaction by the user with the handset supplier can be termed a "touch point," and one where the handset supplier can engage with the user. These touch points can be used to deliver further value to the user, for example through the provision of downloadable content or a smooth upgrade path to the next new handset. From a business model perspective, the longer and more often a user can engage with a mobile handset, the more opportunities there are to monetize this behavior and create further customer loyalty.

In Figure 2.1 we have defined a framework of eight factors that influence the design of a mobile handset and contribute to the central core engagement value. These eight factors are: trends, desirability, user journey, usability, business models, content, applications and services, and the Internet.

These factors are not in any particular order; however, they are all inter-related, e.g. an Apple iPhone may be desired by some people due to the easy usability and choice of a large number of downloadable applications. This combines usability and applications as two influences, which, if done well, increases the engagement value of the handset design.

Not all the factors may need to be as highly prioritized when designing a mobile handset for a particular market. For example, in certain market segments users may not need the ability to browse the Internet on their handsets, whereas this will be very important in some other markets.

An explanation of each factor is given throughout this chapter, as well as how each factor can be linked together with other factors within the design influences framework.

We might question why "technology" is absent as a first-level factor in influencing the design of a handset. Technology is often a key enabler for value creation in a handset design; however, it is not a provider of that value in and of itself. If a handset is being designed to give an excellent gaming experience, it is foremost the desirable experiences of gaming that need to be defined and articulated, after which the type of technology that can enable and deliver this gaming experience can be identified and utilized. For example, the great user interaction experienced on a gaming platform that gives you the feeling of being actually on a skateboard should drive the selection of technology such as motion sensors and actuators to deliver that experience.

Advances in technology make many new things possible, and these advances have been instrumental in how phones have developed over the decades. However, people don't buy technology, they buy experiences. The challenge for designers is to use their skills to work with the raw materials of new technologies to create compelling new user experiences. For these reasons, we have shown technology in the diagram, not as a key design influence, but wrapped around engagement value as a (key) enabler.

2.2 Trends

Trend analysis is a fascinating discipline which involves developing a deep understanding of the macro-economic and social trends and how these affect the needs and desires of the target users of the handset design. Insights from trend analysis form a very important input for designers, in order to ensure that a design is very much in tune with the social demographics, lifestyle and environment of the target audience.

Examples of two mega-trends are the green economy and the aging population. The number of people aged 60 and over is expected to rise from a base of 600 million in the year 2000 to 1.2 billion in 2025 to two billion in 2050. As citizens enjoy improving life spans, governments in many parts of the world are under increasing pressure to manage national health expenditure more effectively for their citizens, especially in the later stages of life. This trend creates a very large opportunity to provide mobile handsets for the over 60s that incorporate health monitoring applications such as blood pressure monitoring or assisting the correct reading of medicine packaging at home. From a mobile handset design perspective, incorporating inclusive design techniques such as larger, easier to read displays, input technologies to aid accessibility and much simpler and easy to use interfaces will help create compelling "engagement value" for this growing market. In response to this trend, there have been several new mobile handset companies specifically created to address this market opportunity.

For the mega-trend based around the green economy, aspects such as extending battery life and using sustainable materials are key factors which will continue to drive handset design. An example is the development of ultra-low-power display technologies for mobile phones and e-book readers, from companies such as Qualcomm and Liquavista (since purchased by Samsung).

Trend analysts within mobile handset companies go into much more detail on the actual micro-trends themselves, from which they derive visual trends and other clues. This analysis informs the choice of materials, colors and aesthetics of the handset design, e.g. use of leather on the handset and accessories. The influence of fields such as art, music, fashion, politics, architecture, science, geography and others all feeds into influences on the handset design.

When Nokia opened up a design office in Brazil, one of the first graphical design projects they undertook was in Rio. The team examined a range of influences from street art, street living and graffiti through to more formal billboard advertising and promotion. It is likely that insights from this work influenced visual elements in mobile devices such as color and material of the handsets, wallpapers or screen savers, or areas like packaging and marketing. Looking beyond city life allows the consideration of how to design mobile devices or services to meet the needs of more rural communities in Brazil and other countries around the world.

In the handset design process, the product roadmap plan needs to be carefully aligned to meet the trends that are observed or likely to appear over the next few years and beyond.

Before a decision is made by a company to develop and launch a handset, a market analysis is carried out to identify where there may be particular existing market niches or brand new markets to be created. A roadmap plan is then created for potential new handsets which takes into account the development times of mobile handsets. The typical development time for a mobile handset is anywhere between six and eighteen months. The development time depends on the complexity of the design, the technology used and the competitive landscape. For example, Samsung and LG have been able to develop, launch and sell handsets for their domestic local Korean markets within six months, whereas Motorola and Nokia may have taken up to eighteen months for a global best seller. Overall, the world-leading handset manufacturers typically have roadmaps looking out between two years to five years ahead for handset product releases. Some of these companies may include research and development teams, looking at five to fifty years in the future to see how the world may look and hence anticipate the impact on the handset design. Using trend research for mobile handset design is a vital part of the handset design process, yet it is also potentially the least accurate driver of handset design, since no one really knows what the future is going to look like in five to ten years time. For this reason, mega-trends and observations of what is going on around the world at a macro level are used, and recommendations based on this are fed into the handset design criteria.

2.3 Desirability

Just like designers of handbags and cars, eyewear and interior design, mobile handset designers follow a common design philosophy of introducing elements of desirability into their "blank canvas" with the hope of delivering an emotional and enjoyable connection with the user. Typical elements of desirability in the mobile handset world are usually centered around visual design, material feelings and brand language association. For example, if Louis Vuitton were to create a mobile handset, key desirable elements would be an extremely elegant design, finest choice of leather and seamless precision of hinges in the handset. To go even further, a consumer of the Louis Vuitton handset would already have a preconception of the brand values associated with Louis Vuitton such as fine handcraftsmanship, high quality and great attention to detail, and so would expect this in the mobile handset. For new brands entering the mobile industry for the first time, it is really important to keep a consistent "design language" of handsets they design so that the consumer knows what to expect and can remain loyal to that brand.

A good example of the use of consistent design language across multiple products is Apple, evidenced by their entry into the mobile handset market with the first iPhone. We believe the Apple brand represents design values such as simplicity, ease of use, crisp and clean design with a minimalistic touch of luxury and transparent technology in the background. Users of an Apple notebook, music player and mobile phone experience desire for these products and remain loyal to the brand. Apple uses these attributes to keep users excited about launches of future product offerings.

Desirability can also extend to the retail outlet. The place where you purchase a mobile handset can be seen as the last point in the value chain of the handset product lifecycle by product designers. Therefore design elements need to be built into the handset at a very early stage to suit the target retail environment – well in advance of the time of purchase. When Vertu, a pioneering luxury handset company created by Nokia, entered the market, they chose established luxury outlets such as jewelers to sell their handsets. Vertu designers had to take into consideration how the handsets were packaged and protected when leaving the manufacturing site and how they would be displayed at the luxury retail store, in order to ensure that no scratches were present to discourage potential customers. These factors have a direct influence on the handset design – from a

€10 000 handset made with expensive materials to a very low cost, basic "throw away" one.

Sometimes desirability can be introduced into the mobile handset design process through strong leadership and a culture based around challenging the normal design rules. The Motorola Razr, launched in Q3 2004, became the best selling clamshell phone in the world and is said to have had enormous design and development challenges, design risks and inter-departmental push-backs. However, there was a consistent drive and support from the top to pursue such a thin and radical design, and the results were over 130 million unit sales. Motorola subsequently tried to replicate the success of the Razr with other handsets by following a similar design approach, but they were much less successful. According to the popular press, some of the reasons for the absence of follow on major successes have been the lack of focus on the user interface design, relatively slower time to market compared to their competitors (such as Samsung and LG Electronics) and not having industrial design as a core design criterion within senior management decision making. Having a mix of people with differing cultural views and tastes, as well as a strong focus on design and creativity, needs to be embedded right at the beginning of the design process and right at the top of the organization to increase the chances of consistently producing global best sellers.

2.4 User journey

Mobile handset designers create a storyboard of typical journeys that the user of their handset will experience and then design in key features, applications and services that will help the user along their journey. For example, if the user is an international business person, then the device will likely have a very good e-mail client, be multi-band, allowing it to operate in multiple countries, and have fast Internet connectivity and browsing facilities. Beyond the applications and services, the journey of how the user engages with the handset retailer, network carrier and the point of disposal of the handset are all identified as areas of differentiation and improvement.

Traditionally, handset designers have designed business-focused handsets around business applications such as e-mails and accessing documents. However, designers have come to realize that business people have other lives during business hours. It is fascinating and insightful whilst sitting in an airport waiting for an airplane to observe a business traveler showing a photo of their newly born baby to their colleagues or playing a game on their mobile. In addition, of course, they may also be checking and responding to their e-mails to keep ahead of the corporate game. The same person, with the same handset, might take on the role of a mother, fashion icon, social networker or business woman at different times of the day. "Situational adaptive user requirements" is the term used to refer to the need to provide the right service at the right time in the right place.

Situational user requirements can be shown as a user journey – for example, showing a business user accessing the *Financial Times* online through their handset in the morning before they enter a breakfast meeting. According to a November 2011 article on TechCrunch, a leading web publication company, the *Financial Times* desktop site is most popular during midday, whilst the mobile version of the site receives a huge spike in traffic between 7 and 8am, presumably from people accessing the site when they first wake up or during their daily commute. Having the ability to download specific material about the agenda of a business meeting, or to read the profiles of fellow attendees on social networking sites such as LinkedIn, can prove to be a valuable resource ahead of walking into the meeting. With the widespread availability of apps on smartphones, certain handset manufacturers are bundling popular applications with their handset software as a point of competitive differentiation and added value.

Mobile handset designers are helping to create a more engaging and valuable journey for the user by working closely with a wide range of organizations outside of the traditional value chain, but who also touch the user. An example would be lifestyle service companies such as restaurants, bars and nightclubs who may wish to provide relevant information and offers through location-based services. This can be an attractive business model for advertisers, allowing them to provide valuable information to consumers at the right time. For example, a group of friends may have just come out of a nightclub in a new city and are alerted by a text message to a nearby restaurant offering discounts to people from that specific nightclub who like jazz and tequila.

Other examples of how the user journey can be translated into the design of the handset include the incorporation of a one-click panic call button for a child in an emergency situation, through to a one-click call button for an international traveler's concierge service to help them find the best luxury hotel near a beach that serves fresh oysters. This use of a one-button call to action can either be hardwired into a permanent button onto the side or back of a handset, or exist as a soft button on a touchscreen which varies depending on the market/user scenario.

2.5 Usability

The number one usability factor that influences the design of a mobile handset is you! The ways you engage, touch, feel, call, text, purchase, download, view the screen, access the Internet and even send e-mails are already documented as a potential "user profile and journey" in a product marketing plan of a handset manufacturer. Increasingly, mobile handset manufacturers in collaboration with mobile carriers and retailers are creating deeper levels of mobile user profile segmentations to categorize the multiple groups a typical mobile user belongs to. For example, the same person may be a corporate user, a luxury user, a social media user and a mother who needs to drop her children at school.

Acquiring a deep understanding of the user happens right at the beginning of the creation of a new handset design, and is a key responsibility of the handset brand. Typically this is the domain of the product marketing/portfolio team within the company, who work increasingly with experts in the field of human–computer interaction. Information is gathered on global mega-trends, taking into account a wide range of influences, for example the impact of music, science, art, society, fashion, culture and environment. These influences are then matched to the target user profile. Descriptions, called personas, of typical target users are created, which help designers to visualize their target users. Narratives, or "stories," are created which describe the user segments to be served and seek to identify the best way to engage with these users. These personas can help to give a much clearer idea of the typical users the handset manufacturer is seeking to serve with their handset. This forms the initial answer to the question "Who is this handset designed for?" The next big question to address is "Why design a handset for them specifically?"

According to Nielsen, a leading market research firm, Apple's iPhone represented 28.6% of the smartphone market by operating system share in Q3, 2011. Considering that Apple were a new entrant to the mobile handset market in 2007, competing with well established smartphone suppliers already in the marketplace, this is impressive. Apple's strategy has been to create and iterate essentially one handset design which is attractive to a wide base of users, across a number of lifestyle user segments. Different segment needs, which traditionally would have been met by a range of handset models, have been achieved through the availability of large numbers of apps, services and content, whilst allowing Apple to provide consistent value to a loyal and growing customer base. Elegant style, simplicity in design and use and an iconic brand statement that reflects one's own brand aspirations have been designed into the handset. In addition, the ability for the handset via its adaptive user interface to give the user what they want, when they want it, at any time is very compelling.

We believe that "relevancy" – making it easy for the user to achieve the most relevant or useful thing that matters to them right now – is one of the next big "game changers" in mobile handset design. One approach is known as "situational relevancy." For example, imagine you are about to step out of the house and head to a cinema. Your phone knows this and finds you the best route to take and the cheapest train ticket, and then alerts your chosen friend to your destination and provides you with the best promotional deals for dining places nearest to the cinema. This combines user journey, desirability and business models, and could be offered via a range of business models – perhaps as a free service from an operator, as a paid for application from a third party, or through advertising-funded models. The delivery mechanism could be via the mobile Internet or simple text messages. You, the user, are in complete control. The key point is that the use of the mobile handset extends into your personal area network in a live manner, giving you instant access on the move. You decide how relevant this is and whether the value being provided is sufficient to keep you engaged.

At the hub of all good usability is a well-designed user interface which takes into account the way a user navigates, inputs commands and uses voice and data on their mobile handset. If we look back at how the majority of mobile users used their devices at the turn of the century, the handset was used mainly for voice and then increasingly text. At the time of writing (2012), use of e-mail and instant messaging (IM) has become very common, especially on smartphones. Usage has grown from business-only increasingly into personal use. It is now a common occurrence to see a fellow passenger on the train or at the airport watching last night's BBC News on an iPlayer over WiFi on their mobile handset and perhaps sending a link of the news report to a friend. This change in usage behavior has a direct influence on fundamental design considerations such as whether to use a Qwerty keyboard, a regular mobile input keyboard or a touchscreen user interface on the handset.

As the processing power for devices such as smartphones increases, so do the capabilities for handling multi-tasking and multi-touch user interfaces. Google Android handsets and Nokia's Windows-based handsets can have multiple tasks on the front screen at the same time. Combine this with multi-touch touchscreen capabilities, and a faster and much more intuitive experience is provided. From a design perspective, incorporating live multi-tasking on the interface affects the decisions made on the choice of chipset, operating system and possible incorporation of a touchscreen.

Today (2012), there are many form-factors emerging for mobile users, especially if one wants to access the Internet using a mobile device, and so segmentation continues to play an important role in the decision process. The challenge for an industrial designer of such products is how best to differentiate these devices if they all look similar, as a result of the presence of a large touchscreen. Differentiation can exist, from the type and quality of the graphical user interface used, to the processing and battery power inside the device and access to exclusive content and curating of content in the apps store. It is worth us looking a little more closely at the choice of the target segment, for example whether they are a corporate user or a personal user.

2.6 Corporate user versus personal user

Careful consideration from the maker and supplier of the handset needs to be given to who actually pays for the handset, for example is it the end user, or is it the company where they work? The answer can have a significant impact on the chosen business model. The individual user may not be the one who buys the product if it is for corporate use. Rather, the purchaser will probably be the IT department of the company they work for, and so different business model criteria may need to be taken into account, such as aftercare services. In addition, corporate IT departments are likely to place a higher priority on requirements such as the need for security of data as opposed to features such as gaming capabilities, although corporate end users who actually use the products often like to play games too!

2.7 Business models

Clearly the amount a user will pay for a phone will strongly influence the features and materials that can be used in its design. However, this is not always straightforward, as many network operators subsidize the price of handsets in order to acquire or retain customers for their network service, with the objective of maximizing revenue from network traffic.

According to research carried out in April 2011 on smartphone consumer purchasing decisions by *Business Insider*, a USA-based publication, "features" were at the top of users' criteria and "carrier offerings" at the bottom, as shown below:

features (most required in purchasing decision);
app selection;
platform;
price;

easy data migration from current phone;
my carrier offers this (least required in purchasing decision).

The nature of network operators' business models mean that they are much more interested in selling airtime contracts than the handsets themselves. More pointedly, the handsets that are sold through retail stores are chosen based on their perceived ability either to increase network usage (for voice and text) or encourage consumers to upgrade to a more expensive bundle of voice, text and data. In many parts of the world, handsets are subsidized by the carriers, which makes it much easier for the consumer to purchase handsets at relatively low cost.

In Japan and Korea the availability of high data bandwidth networks means that many more applications and services are available on handsets.

A more recent trend is for network operators to share network infrastructure and procurement of handsets, in order to seek to lower their capital expenditure costs on both infrastructure and device equipment. In April 2011, Deutsche Telekom and France Telecom agreed a joint procurement venture in an effort to save 400 to 900 million euros over the subsequent three years. From the perspective of the handset brand, if initiatives such as this one create a larger market for a particular handset design, then development costs can be consolidated, better volume prices can be achieved for material purchasing, and better network inter-operability can be achieved. Other attempts to reduce market fragmentation may open up the opportunity for mobile operators and handset companies to enter into production and increase their market share of "ultra-low-cost handsets," in emerging markets such as India and Africa, based on low-cost standard hardware platforms.

How value is monetized from mobile handsets is becoming more important, yet also more confusing, due to a convergence of traditional business models. Historically, it was quite simple, with a branded handset manufacturer making money from handset sales through operator retail stores. The operators would make money on services. Today there is an increase in revenue sharing occurring between the operators, handset brands and third-party application providers. In addition, there is a continued battle for whose brand name appears on the mobile screen. With an increasing shift from voice to data usage on mobile phones, the questions emerging are "Who owns the customer?" and "How does one continually make money from the customer?" For example, consider a user of an HTC Google Android handset built for T-Mobile who chooses to download the popular game Angry Birds. In this instance, is the user's brand loyalty towards HTC, T-Mobile, Google Android or Rovio – the games developer of Angry Birds? Would a user swap handsets or network operator if 20 mobile game downloads were supplied free every month, for instance? It is interesting, for example, that a DVD box set for the movie series *Mission Impossible* was included in the price of a handset offered by one leading UK handset retailer.

In some cases, handset manufacturers seek success of a mobile business venture by focusing on strong brand affiliation of their handsets with the rest of their non-mobile products. For example, a mobile handset from Tag Heuer focuses on excellent design craftsmanship and materials, yet has much less emphasis on specialized content. Point-of-sale retailers are also creating mobile businesses. For example, in the UK, Tesco, a large supermarket chain, has become a mobile virtual network operator (MVNO) providing services on the back of an incumbent operator, and sources some of its handsets directly from Far Eastern original design manufacturers (ODMs) through a local handset distributor.

A major supermarket in the UK has retailed a very thin, credit card size phone designed for party goers which is priced at £18.99, which sold 10 000 units in just two weeks. By contrast, in 2006 a handset containing encrusted diamonds was developed by a Swiss mobile manufacturer called Goldvish, with a retail price of $US1 million. There are different markets, price points and business models depending on the market segment one tries to reach, and hence different handset design considerations to take into account.

2.8 New entrants into the mobile industry

As it becomes easier, cheaper and quicker to make basic mobile handsets, some major lifestyle brands have entered this growing market. Over the past few years, for example, we have seen Porsche, Levi's, Tag Heuer, Armani, Prada, Dior and Ed Hardy enter the mobile handset market. All of these famous brands have done so in partnership with established handset manufacturers, with the design influence mostly coming from the lifestyle brands.

As mobile handsets become ever more embedded into our daily lives, and lifestyle brands look at new ways of connecting with their customers, there will be more brand convergence in the mobile industry. Increasingly, more mobile handsets will display these brands either on the front screen user interface, or through innovative design features of the handsets, and even through being purchased at a brand's traditional retail outlet. It is also likely that there will be an increase in partnership arrangements with branded content suppliers, for example, the HTC HD7 Windows mobile phone with Netflix and Xbox Live on its front screen. Other arrangements include the handset manufacturer or the mobile operator obtaining exclusive content downloads – examples include Apple and the iPhone and a number of operator retail stores. With the launch of the Apple iPad, a small thin tablet personal computer, there is a move away from selling these types of products through operator outlets and towards making them widely available in Apple stores and other retailers of PCs. Equally, operators are already looking to create their own mobile Internet devices, perhaps based on Android, with a form-factor in between the smartphone and the tablet PC. The expectation is that they will sell these devices through their retail stores, with the main revenues achieved through driving data traffic, rather than device sales. The ability of operators to customize the devices, and to personalize the branding through original device manufacturers will provide exciting new price points for cheaper devices with larger screens, more content and fast processors.

2.9 Content

In order to view and experience new content on a mobile handset, one has to ensure that the right software and hardware platform to download and view the content is designed into the handset in the first place.

There is currently (as of 2012) a battle in the mobile marketplace for which smartphone platform can deliver the most applications and exclusive content, and the vibrancy of the developer ecosystem has become a key competitive issue for each of the platforms. The Apple iPhone ecosystem is vigorously managed by Apple with respect to approval of applications, whereas the Android ecosystem is less stringently managed by Google. In both cases, however, hundreds of thousands of apps have been created and can be downloaded. The Android operating system is available for handset OEMs to use within their handsets, whereas Apple's iOS is exclusively available to Apple's own products. At the time of writing, a new wave of products based on Microsoft's platform are entering the market. Ultimately, it will be the end user who decides which mobile handset provides the best engagement value each time they download new content or apps and services. With multiple platforms available acting as delivery systems for the content, apps and services, it is a hard decision for the developers and content owners to decide "which horse to back" in terms of the handsets and which content the user is likely to purchase.

With increasing quantities of user-generated content available through mobile handsets with services such as Facebook, Twitter and photo sharing, one can use a mobile handset and be an author, editor and publisher of content. Some world events have been shown on world media news through the video camera of a mobile phone. This gives enormous power to the users, yet also poses a threat to traditional journalists and content providers. A key issue for the handset manufacturer today is deciding on the role they are going to play in the mobile value chain in the future. Will it be the traditional model of a one-off price for the product, or, for example, could it be via advertising impressions or clicks on pages via the mobile Internet, or by purchasing relevant data and music tracks? All of these issues are highly relevant to the design of a mobile handset, since applications for accessing user-generated social networking sites such as, Flickr, YouTube, Facebook and Twitter can be pre-installed on mobile handsets to provide quick and easy access and with the ability to upload and download new data intuitively. The choice of display technology and the quality and number of cameras play an important role in the handset design process, as these features affect the way in which the handset is used.

2.10 Applications and services

By the end of 2014, Gartner, a leading research company, forecast that over 185 billion applications will have been downloaded from mobile app stores since the launch of the first app store in July 2008.[1]

The growth in the use of downloadable applications has implications on the choice of processor, displays, sound system, on-board sensors, software, hardware platforms and input technologies, as manufacturers seek to enhance the use of these applications and provide differentiation.

Examples of some common applications today are virtual pets and farms, motion-sensing games, creating Koi fish on your mobile handset, and accessing the radio. One phenomenally successful gaming application, Angry Birds, developed by Rovio Mobile in Finland, has been downloaded over 500 million times.

With the availability of increasing data bandwidth and the emergence of location-based services, mobile TV and mobile gaming, further complexity is added, and yet also new opportunities are available, for more dedicated and personalized handsets.

Having more content and services on the handset drives the need for larger format, which in turn drives the rise in mobile Internet devices that are the size of paperback books and priced at a similar price to smartphones. Examples include the iPad from Apple, the Kindle from Amazon and a wave of Android-based tablet devices from numerous manufacturers.

Another force driving innovation in handset design is from the growth in emerging markets. Nokia Life Tools comprises a range of services provided in several handsets in India that includes agricultural and educational services for consumers in small towns and rural areas. Nokia Life Tools Agriculture services aim to plug the information gap and fulfill the needs of farmers, by providing information on seeds, fertilizers,

[1] Source: Gartner, January 26, 2011.

pesticides, market prices and weather via mobile phones. These services are also being considered for use in other developing countries. It is worth noting that applications and services can be personalized for local as well as global needs.

2.11 Internet

According to IDC, a market research organization[2] "By 2015, more U.S. Internet users will access the Internet through mobile devices than through PCs or other wireline devices."

From a handset design perspective, changing usage patterns have a major impact, ranging from the choice of the mobile Internet browser to the choice of radio bands and mobile broadband data requirements, be they 2.5G, 3G, 4G or beyond. Additionally, the ability to surf the Internet on an easy to use and see display surface is already giving rise to several multi-form-factor Internet-enabled devices, such as large tablets and smaller book-size tablets.

2.12 Innovation in handset design

Innovation may come not only from new technologies, but also from simplification of existing technology, for example undertaking mobile payments via text messages. Innovation may also originate through looking at other sectors, for example mobile heart rate monitoring applications. A range of energy- or environmental-related issues are currently rising in importance, including wireless charging without cables, low-power displays, better designed antennas and the adoption of a universal "micro-USB" to avoid multiple power cables.

Innovation based on human-centered design backed by an internal culture focused on good design, risk taking and breaking rules within the company will inevitably produce very successful products; for example, the iPod from Apple was not the *first* MP3 player in the market, yet it

[2] Source: IDC, September 12, 2011.

took into account user delight touch points such as lifestyle association, packaging, simplicity of downloads, the ease with which emotional value can be experienced and, in some cases, expressed outwardly through industrial design.

Sometimes, it is all too easy even for the biggest handset manufacturers to get it wrong when it comes to design influences. For example, Nokia misjudged a change in the market, when a sudden demand for clamshell or folding handsets emerged, initially in Asia, hitting a peak in 2004. Nokia assumed the demand for clamshell designs would be relatively small and that the costs of creating a new platform and addressing the complex mechanical and manufacturing design issues would be high. The view was taken that it was better to retain the original candy bar form-factor. This provided a window of opportunity to companies such as Samsung and LG, and they took the market share. Nokia soon spotted the missed opportunity, and came back with a line of clamshell handsets.

2.13 Legislation

With increasing amounts of data being consumed by smartphones, and potentially very sensitive data such as credit card or medical data being stored in the "cloud" there are rising data protection challenges. For example, if data is stored in one country and accessed by a mobile user in another, and both of these countries have differing data protection legislation, this could lead to data protection infringement.

According to Verizon Communications, a large US phone company, its spend on cloud services and technology will grow from $US2 billion in 2011 to $US90 billion by 2015. Provision of secure data encryption solutions by carriers is expected to increase significantly during this period.

Handset manufacturers are responding with innovative technologies, such as fingerprint authentication and secure protection of data when making mobile payments through near-field communication (NFC).

2.14 Conclusion

Understanding the needs of the target audience is key to creating the engagement value and then creating and building all of the blocks to provide maximum value to the user. It is important to engage with the target audience in order to see what is missing from their lives or what can be done better. One point of caution is that "bigger, faster and more of it" may not always be the right thing. For example, significant investment has been put into launching 3G services to consumers in India, yet there remains a large group of consumers who are content to hear the radio on their current 2G handsets and use voice and text services.

As new market segments continue to be created to cater for the changing needs of ever more demanding consumers, the mobile industry is responding rapidly with hardware and software handset reference platforms that have advanced inbuilt features such as configurable 3D user interfaces and highly usable touchscreen interfaces. These advanced platforms provide handset designers with greater flexibility and shorter timescales whilst requiring fewer development resources. As a result, the costs in creating handset designs that can be personalized for multiple sectors continue to reduce.

With the increase in the use of applications and services through mobile handsets, we believe we will see more "one-button" access to these applications and services. The brands behind the most popular applications and services are already creating handsets around their brand, for example the HTC Status handset, commonly known as the "Facebook" phone. This phone has a dedicated Facebook button and tighter software integration around Facebook status and updates. This emerging form of alliance between the handset manufacturer and the provider of applications and services demonstrates how the key design influences described in this chapter can help differentiate the resulting handset from the rest of the crowd, so focusing on delivering a more personalized end-user engagement.

In order to make sound decisions about the technical design of a mobile handset, including the selection of the most appropriate platform and components of a handset, and how they fit together, it is important to have a good appreciation of the architecture of a typical mobile handset. In the following chapter, we examine the architecture of a mobile handset, considering this from a number of different perspectives. In Chapters 4 and 5 we examine the design issues for all major hardware and software components, before returning, in Chapter 6, to the design process required to move a handset design from concept through to production.

3 Design architecture

Having examined the factors which influence the essence of handset product design and what distinguishes one handset design from another, it is now appropriate to "lift the bonnet." We need to gain a good appreciation of what the major component parts of a handset are, how they are put together, and what the major design considerations are. This exploration will expose us to a range of high-level technical design factors, which are unpacked in further detail in the hardware and software component chapters that follow. Our goal is to continue to develop understanding of the major design decisions and challenges faced by handset designers, which result from both rapid advances in technology and the equally rapid advances in market opportunities and expectations.

3.1 Design perspective

Immediately we are faced with dilemmas over how to describe the structure and organization of a mobile handset. Depending on where we sit in the design flow, what question we are trying to answer, and which information we are seeking to understand, we will view the internal structure and organization of a mobile handset design very differently. This is because a modern mobile handset is probably the world's most complicated consumer electronics product, and contains the combined efforts of many thousands of people over tens of years. It is possible for it to be so complex, at such a relatively low price, due to the very high volumes of sales achieved, allowing the costs of R&D and manufacture to be shared out (amortized) across millions of consumers who buy the products.

Think of studying mobile handset design as one would study the world through the pages of a "world atlas." With our atlas we can understand the world in different ways. We could view a map of the physical characteristics of the world, showing us mountains and rivers and deserts and so on. Equally, we could view a map of the world which showed us the different political boundaries which make up countries of the world, perhaps with their major cities shown. Two different views of the world: both correct, but different in the perspectives they bring.

So, with understanding how a mobile handset works and is designed, depending on which aspect of design we are trying to understand, we need to select the most appropriate map. In this section we are about to take six different perspectives on the structure, or architecture, of a mobile handset as follows:

- from a product teardown point of view (looking at all of the individual components which constitute the handset);
- from a standardization body's point of view;
- from a hardware structure point of view;
- from a software structure point of view;
- from a mobile handset manufacturer's point of view;
- from a network operator's point of view.

We trust that, having examined what constitutes a handset from a number of points of view, you gain a better appreciation of the key design issues at a conceptual level. We will leave it as an exercise for the reader to consider a mobile handset from the point of view of the end user. As a wise person once said: "You can only view one side of a mountain at a time. See all that you can from one perspective, but know that it is never all."

3.1.1 Platform approach

Throughout this book, reference will be made to platforms and "taking a platform approach." A definition is called for before we proceed much further. A platform shall be defined as an integrated set of components that enables a defined set of finished products (in this case mobile handsets) to be created in an economically efficient manner, whilst allowing an appropriate level of market differentiation or customization to be provided by the partners and customers of the platform provider. By way of illustration, the chip manufacturer Intel defines on its website its platform approach as follows: "An integrated set of ingredients that enables targeted usage models, grows existing markets and creates new markets, and which delivers greater end user benefits than the sum of its parts."

A platform could, for example, be made entirely of software components, of hardware components, or a combination of both, optionally with other service aspects such as customization packs, training, support services and a software development kit (SDK), all provided to support market differentiation.

Platforms, if done right, provide great benefits in terms of time to market (less integration and testing for each customer to undertake themselves) and cost (volumes of common components can be aggregated across multiple customers), and, with an appropriate level of customization possible, allow vendors such as OEMs to focus on those aspects of product design which make them stand out from their competitors.

3.2 Physical view – a product teardown

The first point of view we shall take is the physical view. One very natural and intuitive approach we can employ, if we have some knowledge of modern consumer electronics, is to buy a mobile handset from a retail outlet and physically start to disassemble the product into its constituent pieces, to see what is inside. This would provide us with a very physical and tangible view of the anatomy of a mobile handset, although such an approach would be unlikely to help us understand very much about how the product actually works.This approach is known in the industry as a "teardown." A teardown allows an "outside looking in" understanding of the constituent parts and likely cost of manufacturing of a particular handset design or other consumer electronics product. Teardowns are typically undertaken by specialist organizations, and normally start from a product which is commercially available. Teardowns literally involve breaking down an existing phone, or other product, into its component pieces. Such organizations have gained considerable experience in judging the likely purchase price of the components that make up the product. In some cases, custom or semi-custom chips may have been created for the product, in which case the teardown analyst will seek to infer the likely cost point of a component based on what the chipset is likely to achieve, its size, packaging and so on. It is worth noting that, in addition to the services of specialist organizations, many mobile handset companies also undertake their own teardowns – especially since they will know the price that *they* are paying for a number of the components of a competitor's product.

The result of a teardown is a detailed report which itemizes all of the component parts of a mobile handset. Based on detailed industry knowledge, an estimate is made of the cost of each individual component, and, where possible, the name of the supplier of each component. Along with other costs (which we will come to shortly), the teardown can then provide an estimate of the manufactured cost of the handset.

3.2.1 Bill of materials

A list of all of the tangible component parts required to build a product, along with their quantity and prices, is commonly known as a "bill of materials," or BoM (pronounced "bomb"). So the BoM will contain a line item for each physical component part of the handset – chipsets, cameras, display, connectors, sensors, PCB, caseworks, battery, down to individual capacitors and resistors (collectively known informally as "birdseed" due to their small size and relative abundance).

Pricing of components is heavily contingent on the expected volume of products made. Essentially, the higher the volume, the lower the price per unit of a component. The exact price agreed between a component supplier and the handset manufacturer is a closely guarded secret. Some of the factors which influence the price agreed include:

- the investment required to achieve the component design;
- the variable costs of production with increasing volume;
- the anticipated total volume of components required during the lifetime of the handset product, balanced with the level of confidence the supplier has in the ability of the OEM to achieve those volumes;
- the strategic value and credit worthiness of the OEM to the supplier.

In Chapter 6, "Product design," we will explore the merits of an OEM taking a platform approach to the development of a range of products, in order, for example, to improve time to market and reduce incremental R&D investment. For now, it is worth noting that if the same component can be used across multiple products, then, as the total volume of that component required rises, the average unit cost of the component will trend downwards, and this should lead to a reduced component price point, leading to a lower unit cost for the final OEM products in the marketplace.

3.2.2 Manufacturing cost

In order to arrive at the final estimated cost of a mobile handset, a number of other factors need to be taken into account. It is all very well having a "bag of bits" which are the physical components of the product, but there are numerous other cost-incurring elements to take into account. The most obvious of these is the manufacturing cost required to assemble, test and package the product ready for distribution.

Manufacturing cost is often initially estimated as a percentage of the BoM. Although this seems a coarse measure, it does give a good first approximation to the complexity involved in manufacturing a product. A product with a low BoM is, generally speaking, going to be a simpler product to manufacture, and a product with a higher BoM will be more complex, or more novel, and therefore likely to be more expensive to manufacture. Just as with the cost of components, manufacturing cost tends to be less expensive per unit with higher volumes, due to the very high levels of automation in production. For the manufacturer, a similar set of criteria to the component supplier comes into play in negotiating a final price with the OEM – new investment required to manufacture the product, anticipated volumes, OEM credibility, credit worthiness and the strategic value of the account to the manufacturer.

3.2.3 Mechanical components

Another factor included in the manufacturing cost is the cost of the physical molds for mechanical design elements, such as caseworks and housing structures to hold components such as the battery in place.

Custom jigs and other fixtures also have to be designed, for example to hold the PCB in place during factory testing. If it is a new product range, then bespoke factory test software may also need to be developed to allow automatic testing of the product through software control, and to program the software code image into the flash memory of the product. Typically these "one-off costs" are amortized over the volume expectation of the product. Hence a casework mold costing $US500 000 will be amortized to 50 cents over a product run of one million units. It is worth noting here that molds have a limited lifetime, and typically need replacing after a production run of approximately two million units.

3.2.4 Accessories and packaging

Other costs include material costs for accessories and packaging – including, as a likely minimum, battery, charger, headset, manual, box materials and a cardboard sleeve. In addition, there could be a serial cable, PC software CD, a memory card, and possibly the SIM card for the specific operator customer. Combining these costs allows us to compute the *ex-works* price of the product – basically the unit cost to get a handset to the dispatch point in the factory, but before the OEM's cost overheads and profit margin are added to the product cost.

Once the manufacturing costs have been fully considered and amortized, then a price known as the ex-works or factory selling price (FSP) can be computed. This is the manufacturer's per-unit cost for the creation of a product that is boxed and ready to ship.

3.2.5 Software bill of materials

In addition to the physical components and the very visible building and testing of the consumer electronics product, a modern handset is likely to contain a range of software items from specialist software companies. Typically, this could include the operating system, the browser and specialized functions such as voice recognition. Third-party software is generally priced just like an electronic component, with a per-unit price, which may be discounted as the volume of mobile handsets manufactured increases. In addition to per-unit costs, other expenditure, such as that relating to integration and maintenance, would generally be considered to be under the remit of R&D, and would be accounted for as such.

In the early days of 2G handsets (in the mid 1990s), specialist software consultancies were in a strong position to provide their services to OEMs, who were struggling with the rapid increase in the quantity and complexity of software required for the new generation of digital technology. As OEMs recognized the strategic importance of software and invested considerably in their own software R&D, third-party software companies responded by either putting more focus on providing bespoke consultancy work around integration, testing and product variations, or creating new businesses to develop high-value phone software which could then be licensed to multiple OEMs.

For those handset manufacturers unable to make the significant investment required in software for 2G phones, a small number of specialist software companies emerged to develop and supply the core communication stacks, typically in partnership with a silicon manufacturer. Examples included Condat (later acquired by TI), Optimay (later acquired by Lucent) and TTPCom (the remaining independent supplier, eventually acquired by Motorola).

From the mid 1990s onwards, as third-party software companies began to create modules of software which could be integrated into multiple handset products across many different handset companies, the model of treating software as a component in the cost structure of a handset emerged. This provided an upside for third-party companies if sales of handset product containing their technology were large, whilst encouraging a risk-sharing approach at the R&D stage (little or no revenue) which mirrored more accurately the risk–reward business model of mobile handset manufacturers. All that being true, high-volume manufacturers with the need to rely on a particular software component often consider it to their advantage to return to the "old fashioned" non-recurring expense model, at least in part, to reduce their overall spend in the mid-term on third-party software. Classic examples of early third-party software include predictive text software from Tegic (now part of Nuance), WAP technology from Unwired Planet (long since re-branded Openwave), Java technology from Sun Microsystems or Esmertec and voice recognition technology (also from Nuance, based in part on earlier acquisitions of technology from Voice Signal and Advanced Recognition Technologies (ART)).

3.2.6 Intellectual property

As well as a handset consisting of numerous tangible mechanical, hardware and software components, also embodied in a handset design are numerous inventions and creative works of other companies or individuals, who have made significant contributions to the development and evolution of mobile handset technology throughout its history. Because intellectual property embodies an important, albeit less physically tangible, aspect of the makeup of a mobile handset, we need to address this topic here in brief.

The established mechanism for embodying inventions is through a system of intellectual property patenting, operated on a national basis. Patenting allows the assigning of ownership of an invention, and is a system designed to encourage innovation by providing a legal system to defend against the use of an invention by third parties without permission. Without the patent system, the business incentives for companies or individuals to make investments in research and innovation would be significantly reduced, and the pace of innovation would falter.

Creative works (such as any form of media, industrial designs, written works, etc.) are subject to a system of copyrighting. Copyrighting allows the assignment of ownership of a creative work to the copyright owner, and provides a legal system to defend against the copying of works by third parties without permission.

A further important trend is in open source software licensing, which adopts an approach that actively encourages the copying, re-use and modification of open source software in return for contributions of improvements and, in some cases, other software which uses that open source software, back into a community of available software. Open source software is typically at no-cost at the point of use; however, costs are incurred in a number of different ways, such as in support costs and in contributing additional software back into the community. In addition, open source software is still subject to the intellectual property rights of others where it embodies their patents.

Holders of intellectual property have a business asset, which they could legally apply to prevent others using their patented or copyrighted work and to prevent copying of their intellectual property. However, more commonly, owners of intellectual property have licensing programs to enable others to make use of their intellectual property on fair, reasonable and non-discriminatory business terms. With this approach, the industry, and in turn end users as a whole, can benefit from the inventions and creative works of others, whilst providing an appropriate remuneration, which, in turn, continues to encourage a wide range of companies to invest in further innovation.

Royalties for intellectual property are typically agreed and computed as a small percentage of the selling price of each handset product sold by a licensee. This selling price is typically the price at which the OEM sells their product, either into a distribution channel or to an operator customer, and is therefore not the same as the final price which the end customer pays in a store. In Section 3.2.7, we will cover the other costs incurred between the OEM and the end customer.

Copyrighted works (images, music, logos, etc.) from third parties are seldom found on mobile handsets, unless they are part of a marketing or advertising arrangement which has a marketing benefit to the OEM. A mobile handset will have a number of sample items of music or video shipped with the product to demonstrate the multimedia capabilities of the handset. Most often, these are out of copyright items (famous "old" segments of classical music) or custom creations by the OEM. If copyrighted works are included, then it is quite likely that there is a co-branding opportunity either for the OEM with the copyright owner, or between the copyright owner and the OEM's operator customer. The rationale for all this is that the OEM does not want to add further costs to the handset in licensing copyrighted works, unless there is a clear upside business benefit to the OEM.

3.2.7 Distribution costs

From the description so far, we might believe that once we have accounted for all of these costs, we have considered all the pricing elements that make up the cost of the consumer electronics product we choose to buy from a retail outlet in the high street. From a design and manufacturing perspective, this is true. However, if we consider for a moment the truly global reach of the consumer markets for mobile handsets, we can see that there are significant challenges, and hence costs, in taking millions of handsets manufactured in one location and getting those products distributed, not just to multiple regions and countries of the world, but also to cities, towns and smaller settlements within a nation or region.

So, now we have a complete handset product, packaged in an operator-specific or market-specific box, but it is still sitting in the factory, in many instances in a different country to that of the point of sale. In a traditional supply chain, product will leave a factory and be dispatched to a wholesaler, who in turn will supply a number of smaller retail outlets. For handsets destined for an operator market, the operators themselves usually act as the wholesaler for their range of stores and other consumer outlets. For handsets destined for an "open market," the wholesaler is likely to be a country-specific organization that has access to the retail outlets of a region. As a very rough rule of thumb, we might expect the wholesaler price to be around 20% more than the OEM selling price of the handset, and in turn the retailer price to be around another 20% more than the wholesale price – before government import and retail taxes.

In a number of countries, there may be very high import duties on goods such as mobile handsets. As well as being a revenue generator, these taxes, along with other incentives, may encourage a large mobile handset brand to establish local manufacture of their product; this has occurred in a number of geographies, including in recent years India and Brazil.

As an illustrative example, in Table 3.1 we provide a summary spreadsheet of the cost breakdown of one particular mobile handset – the iPhone 4S – based on preliminary data available publicly in October

Table 3.1. *Cost breakdown of a mobile handset*

Component	Cost (US$)
Main chipsets	
Wireless chipset (baseband, RF, PA)	23.54
Application chipset	15.00
WLAN / BT / FM / GPS	6.50
Power management	7.20
Memory	
NAND flash (32 GB)	38.40
DRAM (DDR / DDR2)	9.10
Display and touchscreen	
Display	23.00
Touchscreen	14.00
Camera	17.60
User interface and sensors	6.85
Electronic BoM	161.19
Battery	5.90
Mechanical/electro-mechanical (casing, fixings, speakers, mics etc.)	33.00
Box contents (charger etc.)	7.00
Full BoM	207.09
Manufacturing (including customization and test)	8.00
Ex-works factory selling price	215.00
Retail price without contract (no subsidy)	749.00
Retail price with contract (with subsidy)	299.00

2011 from IHS iSuppli. This breakdown indicates the key components which make up the cost of the device, compared with the final retail price. Although Apple is something of a special case, operating as a business with very healthy gross profit margins of 41.4% for Q2 2011, it does illustrate the point very dramatically of the difference between the cost of the components and assembling the phone ($US215), versus the final unsubsidized retail price of the product at launch ($US749). In January 2011, Bloomberg estimated a subsidy of $US400 would be paid by one network operator in the USA, Verizon Wireless, for each iPhone sold with a contract. Bloomberg estimated that this would cost the operator between $US3 billion and $US5 billion in a 12 month period.

3.3 Standards view

The second point of view we shall take is the standards view. Years before a new generation of mobile telecommunications is deployed, a process of creating a set of industry standards that all the main industry players can agree on is undertaken. From the perspective of a standards making body, a mobile handset is known as a "mobile terminal" – that is, a piece of equipment which terminates one end of the communication path of a network (with another terminal at the other end of the communication path). For those of us who "live and breathe" in the handset ecosystem, this is quite a humbling perspective – to be pushed to the edges. It is also in many ways an incorrect perspective – the user, for instance, perceives the handset, like themselves, at the center, with the ability to reach out to and connect with others. Ultimately, nearly all revenues in the cellular ecosystem originate from users themselves interacting with other users and with services through their handset, and operator value propositions do well never to forget this, irrespective of how much complexity resides in their network.

Standards bodies are focused on the overall architecture of a telecommunications system. A telecommunications system has many different network components, responsible, for example, for managing the mobility of terminal equipment, routing of traffic (nationally and internationally), interworking between legacy protocol systems, providing the required quality of service and interconnecting with other communication services, particularly these days, the Internet.

The network is designed to support a set of envisaged fundamental services (so-called "teleservices") which the combination of the network and its associated mobile terminals may support. Standards bodies specify how information is transported between mobile terminals and the network (so-called "bearer services") and the protocols (exchanges of information) which are required between the various network elements and the mobile terminal to deliver those services. Because cellular standards bodies are typically working on a 5–10 year future horizon, it is almost impossible to envisage which services will be successful and which will not – therefore the focus is on fundamental issues such as the most efficient use of the network to increase the bandwidth speeds available, whilst ensuring that handset designs can be as power-efficient as possible – by not needing to transmit or receive any more often than strictly necessary.

3.3.1 Mobile terminal plus SIM card equals mobile equipment

Within the 3GPP family of specifications, the mobile terminal is viewed as consisting of two fundamental components. The first is what we would call the mobile handset, known as the mobile equipment, or ME. The second is the SIM card (Subscriber Identity Module – known as a universal SIM or USIM in 3G standards). This distinction between the SIM card (representing the subscriber's identity, including their unique serial number, telephone number(s), short message mailboxes and user settings) and the mobile handset was built into the original GSM specifications in the 1980s. At this time, the view was that the subscriber might well wish to use different mobile terminals over the course of a day or a week, perhaps with different capabilities, and would move their smartcard (SIM) between handsets. Two SIM card formats were defined – a *full size*, which is "credit card sized," and the smaller *mini size*, which is the prevalent size that we all know and use today. They were defined to be electrically identical – with the contact points arranged identically – and the cards were of the same thickness. A few early-generation GSM phones were launched in the early 1990s with the credit card sized SIM card, but it

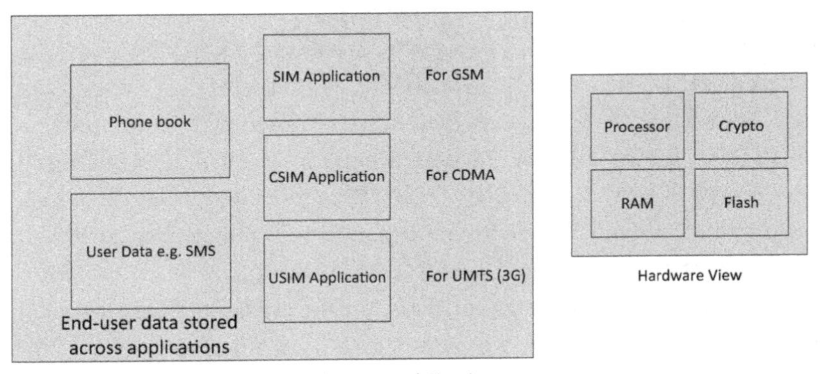

Figure 3.1. SIM architecture – logical and hardware views.

quickly became apparent that there was not a real end user need to swap cards between handsets (handsets were expensive in those days so using more than one phone was very unlikely), and that the smaller SIM card size was much better for engineering smaller handsets. More recently, a third SIM format, the *micro-SIM*, has been defined (also called the *3FF card*), which is 1 cm shorter than the mini-SIM and remains electrically identical to the two other SIM formats. The micro-SIM is designed to fit in devices which would otherwise find the standard mini-SIM too large to accommodate. Curiously though, the first device to support the new micro-SIM format was the Apple iPad – hardly the smallest of devices.

Within the CDMA community, this same vision of SIM roaming has not existed. Hence, historically, most CDMA mobile handsets have not utilized a SIM card, and all of the user information is stored within the handset. This means that there is a one-to-one relationship between the serial number of the mobile handset and the subscriber of the tariff associated with that handset. Under most circumstances, this provides the experience the user expects, although there can be additional administrative steps in migrating a user's subscription from one handset to another. In response to this disparity and to support "world phones," a version of the SIM card for CDMA was later developed known as the CSIM (CDMA SIM) – electrically and mechanically identical to the GSM SIM card, but with some different information stored on the smartcard.

For 3G (or its standards-based name of UMTS), another file structure (known as a *SIM application*) was created, called the USIM. All of these different SIM applications (GSM SIM, CSIM, USIM) can be brought together onto the same physical smartcard if required, and this is known as the R-UIM (Removable User Identity Module). Note that the subscriber's phonebook is stored outside of this SIM application structure, which means that the same phonebook information may be accessed across all the different mobile terminal types (GSM, CDMA, UMTS).

For completeness of our description, the specification of the physical and electrical nature of the smartcard is now known as the UICC (Universal Integrated Circuit Card). Standards also exist for a range of future IP-based multimedia telephony services, and in this case yet another SIM application called the ISIM has been introduced.

3.4 Hardware architecture view

The third point of view we shall take is the hardware view. In this section, we define hardware to mean the physical design of the handset and its components, as perceived once the mechanical housing (e.g. caseworks) of the handset is removed. The hardware of the handset sub-divides into two main categories. The first of these comprises the more mechanical parts of the handset that the user can readily perceive or understand – the display, the camera(s), the keypad (if it's not a touchscreen phone), the antenna, the loudspeaker, the microphone, the buzzer (for alerts), the SIM card, external memory card connectors, serial interface and battery charging connectors, the button battery and its connector, and various backlights for the display and keypad. These together are often known as the peripheral components. The second category includes the printed circuit board (PCB) and the many electronic components mounted on both surfaces – more on this shortly. Many of the peripheral components can be manufactured and supplied as standard across a number of different handset designs – examples would include the audio and memory peripherals and the display. Other components are designed to fit the particular 3D design of the handset – most notably the internal antenna, whose 3D geometry is fundamental to the performance of the antenna, and is an engineering and design discipline in its own right.

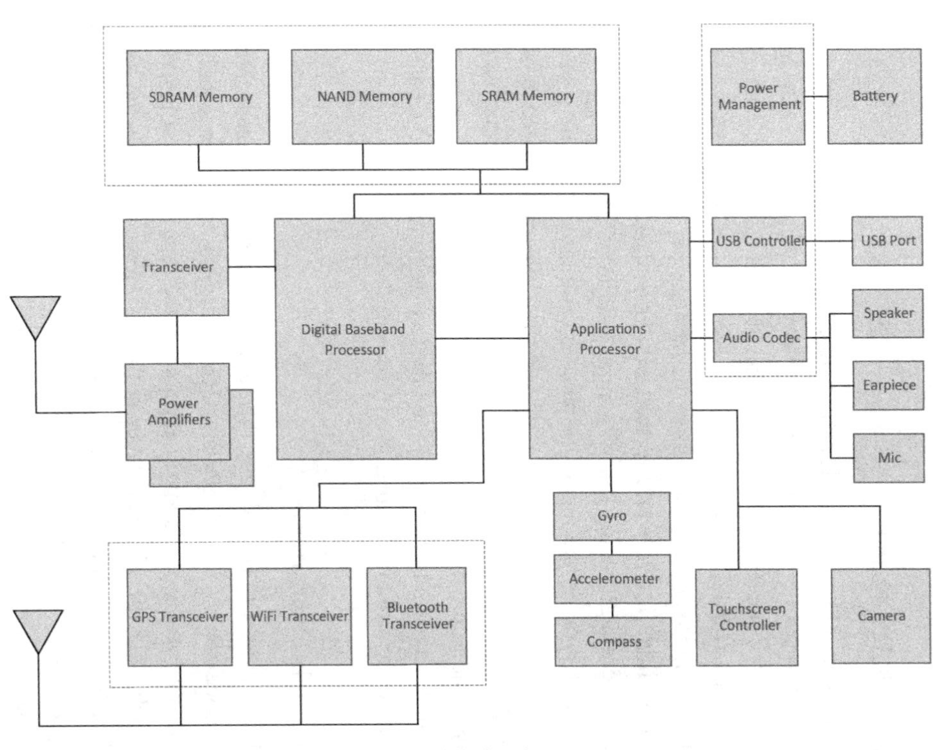

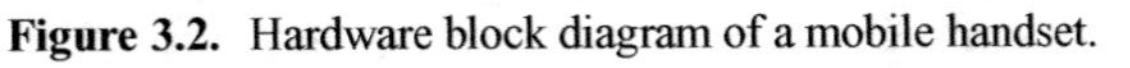

Figure 3.2. Hardware block diagram of a mobile handset.

3.4.1 Printed circuit board

Turning to the PCB design, we shall group together major functions of the handset into digital baseband, application processing, radio (RF), auxiliary modem and memory.

3.4.2 Digital baseband

The digital baseband chipset is responsible for processing all of the communication with the cellular network, based on data signals sent to and received from the air interface via the radio chipset. A modern digital baseband chipset contains multiple processors, both microprocessors and DSP (digital signal processors), which together provide the computing capabilities for software to execute – processing, memory, timing and peripheral interfacing (or I/O – input/output). The signal processing capabilities provide the means to take raw digital information received from the radio subsystem and re-create the high-level sequence of bits which the network has sent to the handset – and vice versa from the handset side to the network side.

Further detail on the functions and the key design issues of a baseband chipset are covered in Chapter 4, "Hardware design."

3.4.3 Application processor

Until the early 2000s, all but the highest-end devices used a single microprocessor core (typically an ARM7 core) to run the protocol software as well as all of the application software required, with a separate DSP to handle the physical layer software. However, with the rise in multimedia capabilities – camera/camcorder, music, video, etc., a tipping point was reached at which the required level of real-time performance could not be achieved with a single processor solution. For a few chipset iterations, improvements were achieved by packaging the low-level camera control and decode features into what was called a companion chip. This type of chip contains discrete logic, rather than a general purpose processor. Some chipset manufacturers also created discrete logic to encode and decode audio and video streams, and built this logic into their baseband chipsets, either as discrete logic or as DSP software. Despite these efforts, a well established trend now is to separate out the "modem" features into one modem processor solution, and the application and multimedia functions into another application processor (AP) solution. Chipsets designed for smartphones – with good graphics and multimedia performance – further split out the multimedia functions from the application functions into a separate graphics processor, or GPU. These multi-processor solutions may exist as separately packaged chipsets – possibly from different chipset suppliers. Increasingly, however, due to the high levels of system integration required to optimize performance "from application to antenna," chipsets (such as Qualcomm's Snapdragon range) are now on the market, which combine the application processor, GPU and modem processor cores onto a single chipset solution. It was interesting to observe that Nvidia – a company noted for its high-performance graphics processors – purchased a wireless modem company (Icera) in 2011, giving it the opportunity to create ever more integrated chipset solutions in the future.

Further advances are being made by the prime supplier of processor cores for mobile devices – ARM – with the introduction in recent years of processor solutions that allow processing functions to be divided between multiple processing cores within a unified processor design. These so-called multi-core processors provide a key advantage of improving processing performance whilst minimizing any increase in power consumption. Rather than working one processor extremely hard, by sharing out the work between processors, less overall power is required, especially if the tasks to be performed can be cleanly partitioned (for example, playing a video clip within a web browser session). As an example of continued advances in application processor core technology, ARM's Cortex A7 processor is expected to offer five times the energy efficiency of the previous generation Cortex A8 processor and yet is one-fifth of the size. ARM have indicated in a news release in 2011 that the A7 will enable smartphones that will cost less than $US100 in 2013 to have the same performance level as those that cost $US500 in 2011.

3.4.4 Radio

The radio section of the handset design is housed in metal shielding in order to reduce significantly electromagnetic interference of the radio from unwanted signals and provide electromagnetic protection for the remainder of the handset, and the user, from the radio's own signals. In a modern radio design, much of the radio functionality is contained within a single radio chipset. Depending on the target market for the resultant handset, the radio chipset will contain support for many different radio standards and operating frequencies.

Current radio chipsets provide support for multiple air interfaces. A typical list of air interfaces which could be supported in the most advanced combined radio chipsets would include:

- UMTS/HSDPA/HSUPA (850, 900, 1900, 2100 MHz);
- GSM/EDGE (850, 900, 1800, 1900 MHz);
- CDMA EV-DO Rev. A (800, 1900 MHz).

Earlier generations of multiple radio support focused on supporting multiple frequency bands for the same radio standard (a band is a contiguous set of frequencies within the radio spectrum). So, for instance, with GSM-only radios, the first radio chipsets supported a single radio band, and then over time added two, three and four other bands – providing a radio chipset which could be used to design a GSM handset which worked in any region of the world with suitable coverage.

With the transition from 2G to 3G networks, it has remained important for network operators to provide legacy support for their customers – such that if 3G network coverage is not available, then the handset can "fall back" to a 2G mode of operation. Early 3G handsets contained two discrete radio chipsets – one for 2G and another for 3G, which was one factor in the high cost of early 3G handsets. Contemporary 3G handsets contain a radio chipset that supports both 3G and 2G, and, at the time of writing, chipsets additionally supporting LTE are now entering the market.

In addition to the radio chipset on the radio section of the handset PCB, the other key component is the power amplifier (PA) chip. The PA is responsible for magnifying the power strength of the RF signal to be transmitted, quickly and accurately, before it reaches the antenna.

This chipset is separate from the radio chipset as PAs have a unique set of design parameters to achieve (for instance) high power output, high efficiency (ensuring as much of the power as possible is used to increase the output power of the signal) and optimum heat dissipation.

3.4.5 Auxiliary modems

A contemporary handset may support a number of other radio standards such as an FM receiver, Bluetooth, WiFi and GPS. Most often, these features are provided via an auxiliary connectivity chip, which is a component that supports all of the non-cellular radio requirements into a single chip. Variations of this chipset can exist, preferably with common packaging and pin layout – allowing the handset designer to add or remove readily a connectivity capability from their handset range, without needing to undertake a major re-design of the PCB or, worse, the mechanical design.

Alternatively, some of these auxiliary modem functions could have their digital communications capabilities built into the digital baseband, with the advantage of reducing the number of chip packages on the PCB, leading to a smaller, perhaps lower cost, design.

3.4.6 Memory

As well as digital chipsets, the other key component on the baseband portion of the handset PCB is memory. Depending on the digital chipset, some memory could be pre-integrated in the chipset design or pre-packaged within the chipset packaging. Nonetheless, in order to support different handset variants with different memory requirements, there are typically still discrete memory components provided separately on the PCB.

Two types of memory are required. The first type is memory to store software program code and static data (data which does not need to change its value – this could include text strings). This memory is known as non-volatile, and requires an electrical current to be applied in order to erase and re-program the memory. Mobile phone memory, along with memory cards, is most often NAND-based, due to the high memory densities that can be achieved.

The second type of memory is RAM, which is used to store temporary data required by the handset software as "working memory," as the software code executes, often transforming data from one format to another (e.g. receiving video content from the air interface and transforming it into a data format which can be written to the display). There are two main forms of RAM available – static RAM (SRAM) and dynamic RAM (DRAM). SRAM is more expensive to produce, but is faster to access and requires less power than DRAM. SRAM is therefore very suitable for cache memory, where speed of access is key, whereas DRAM is more suited for storing larger quantities of data for longer periods of time.

All types of memory are available in different memory sizes. Due to the design and manufacturing processes, these memory sizes are multiples of "sensible base 2 numbers" – such as 1 Mbit, 2 Mbit, 4 Mbit, . . . , 1 Gbit, 2 Gbit, etc. This introduces a constant design trade-off in handset design between fitting the smallest memory size you can – in order to save cost – and fitting the maximum memory size you can – in order to support more functionality and user data. Unfortunately, there have been more than a few occurrences of handset designs that support advanced camera or multimedia capabilities, but have far too small an amount of user memory in order to store photos, music or video. Since the introduction of the microSD memory card form-factor in 2005, external memory cards have entered the mainstream. These small cards, although fiddly, are better suited to inclusion in a mobile handset design, and allow variable amounts of user data to be stored, depending on the user's budget, rather than loading all of the cost of the memory onto the base price of the handset.

Memory manufacturers supply so-called "stacked" memory products. These contain both flash and RAM memory chips stacked on top of each other, within a single package. This is of significant advantage in reducing the "footprint" of the memory package, but does mean that the memory manufacturer must get the choice of *X* amounts of flash and *Y* amounts of RAM right when creating these stacked devices, so as to match the memory needs of handset designers correctly.

3.5 Software architecture view

The fourth point of view we shall take is the software view.

3.5.1 Historical perspective

One of the first things to say about mobile handset software is that it is essential for the same software to be able to work across multiple hardware platforms, and often multiple screen types and keypad layouts. This is necessary for economic reasons – to allow the main functionality of the software (sometimes called the "core software") to be written and tested once, and then used cost-effectively across countless different handset product variants. The smaller modules of software that connect the core software together with other software and hardware components are written specifically for a handset platform. These modules are sometimes called *adaptation layers*, with the more common term *hardware adaptation layer* (HAL) used to describe the point at which core software interfaces to software components responsible for interfacing to hardware. This approach of core software and adaptation software makes handset software more economical to produce – in particular, allowing the majority of effort expended to create, improve and test the core software to be applied once, and permitting a core roadmap of new features to be planned somewhat independently of the vagaries of any particular hardware design or niche market segment.

This approach holds true for both in-house software created within a major silicon platform provider or OEM, as well as third-party software which may be licensed to many different OEMs. The story was different in the early days of the first 2G handset designs. At that time, most handset software was created in-house within the OEM organization, with significant levels of investment of effort required to create the first set of working marketable products. The most significant effort was put into creating the modem software that allows the phone to communicate with the network, with much less effort put into the user interface, which in any event was quite simple by later comparison.

With an emphasis on basic telephony functionality, including receiver performance and power consumption, and with the size of the market for 2G handsets still to be demonstrated, many of the early generations of handset software were often not created in a modular fashion. Development lifecycles were typically longer than two years, and market segmentation consisted primarily of a high-tier, mid-tier and low-tier split.

As phones became more capable, resulting from a combination of ever improving levels of chipset integration, and increasing market volumes and reducing costs, they began to support more and more features, including multimedia and Internet capabilities. The need to support more complex features, along with the introduction of mobile data and the ability to access servers from a mobile device, led to the need to create mobile software platforms which could support a range of different applications that were under development – some by the handset brands themselves, but many others by third-party developers and service providers such as the network operators.

Although there are significant differences in complexity and quantity of software embedded in a smartphone and in an ultra-low-cost handset, there is nonetheless a clearly recognizable overarching structure. An idealized high-level software architecture is outlined in Figure 3.3. We will describe this architecture using the concept of four fundamental software domains of a handset: the *application* domain, the *wireless modem* domain and a *device driver* domain; in addition, there is an *operating system* domain, responsible for the allocation of hardware resources such as memory and processor time. We shall now take a high-level look at the first three of these domains.

3.5.2 Application domain

The *application* software's key functions are to present the handset functions to the user in an easy to use manner, to handle input from the user (the *user interface*), and to provide user control of handset functions – such as making a call, sending a text, taking a photograph or browsing the Internet (the *application functions*).

The application software is responsible for managing the interaction between the user and the functions of the handset. This includes the user interface (what appears on the screen, how the key or touchscreen input

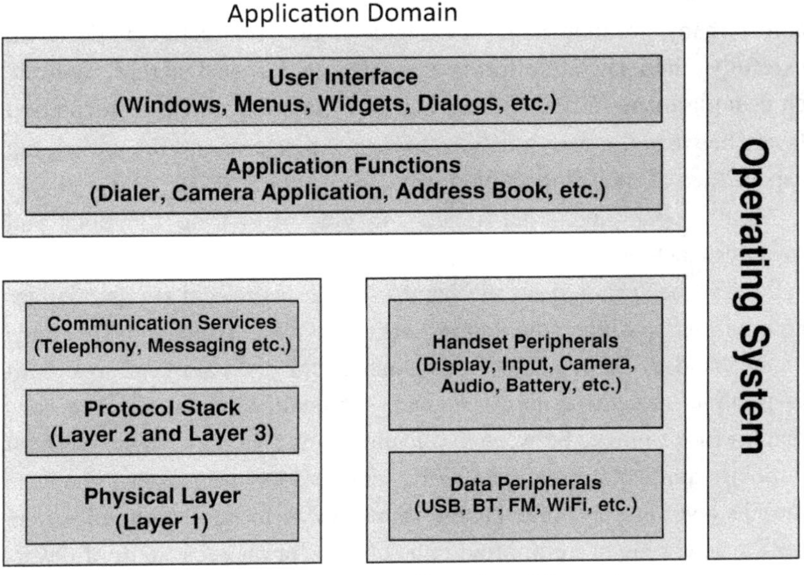

Figure 3.3. High-level architecture of mobile handset software.

allows navigation through use of menus, along with the selection of options and the operation of handset functions) as well as making calls, using the camera, playing a video and so on. For handset functions such as telephony and text messaging, the application software interacts with the wireless modem software. However, for a large range of contemporary functions, such as digital camera, multimedia, Bluetooth, mobile Internet browsers, etc., a number of discrete and specialist applications are needed, which in turn require support for some common functions, such as drawing on the screen, or sending data packets over the mobile Internet.

The need to have a set of shared functions, which may be developed and tested once, and then called by multiple functions or applications, has been the prime driver behind the development of mobile application platforms and mobile operating systems (described further in Chapter 5, "Software design"). Mobile application platforms and operating systems provide for the significant investment in software R&D to be re-used across many different handset designs. If the architecture is designed carefully, then such platforms can also be re-used across multiple chipset platforms. This can be achieved by creating software adaptation layers between the core functions of such a platform and the underlying capabilities of each chipset platform.

Mobile operating systems take the idea of sharing functions and resources between applications to the next level, by enabling applications to be installed on the handset in a secure and robust manner post-purchase, whilst also gaining access to all of the shared functions of the handset. Management of resources for each application – both to provide enough resources to each application and to resolve contention for resources between applications – is a key difference between a mobile application platform and a mobile operating system. Because mobile operating systems enable third parties to develop applications for the operating system, providing support for an ecosystem of developers becomes a key cost in developing and maintaining an operating system.

3.5.3 Wireless modem domain

The *wireless modem* software's key functions are to support the establishment and management of different types of communication services for circuit-switched and packet-switched communications (the *communication services*). This is achieved by the transfer of signaling messages between the handset and the network (the *layer 2–layer 3 protocol stack*), taking into account a wide range of mobility issues, whilst encoding and decoding signaling and traffic data via the physical medium of the radio spectrum through the control of the radio subsystem (*physical layer*).

The physical layer software is primarily responsible for the transmission and reception of digital binary information (bits) between the handset and a cellular base station, by translation of those bits to and from RF signals which are propagated through the RF spectrum. In order to achieve this task, the physical layer software must orchestrate and control a range of hardware components such as the RF transmitter and the RF receiver. It must provide functions to encode and decode the bits to ensure that the original sequence of bits can be reconstructed at the other end, given that the RF medium may suffer from interference and data loss. It must carefully maintain synchronization in time with the network, and must manage timings to ensure that bits are encoded and transmitted at an expected time, and that it attempts to receive and decode bits at the correct time. It must adjust the transmitter power to optimize the signal strength and adjust the receiver gain to optimize its ability to receive successfully. All of these functions must be managed so as to minimize the use of power, in order to maximize the life of the battery between charges.

Physical layer software is divided between functions which are performed on a microprocessor and those which are performed on a digital signal processor (DSP) – see Chapter 4, "Hardware design," for a definition of the difference between the two types of processor. Both of these processors are hardware components of the baseband chipset, which forms the heart of the telephony capability of a handset.

The protocol stack software is responsible for a set of communication dialogs between the handset and the cellular network to achieve a wide range of different tasks, such as registering the handset to a particular network, authenticating the handset to the network, reporting where the handset is (so that the network can route incoming calls to the handset), managing the handover of communications from one cell site to another as the handset changes its physical location, managing the process of setting up and closing down telephone calls, packet-data transfer and the transmission and reception of other information such as text messages (SMS).

3.5.4 Device driver domain

The *device driver* software's key functions are to operate hardware devices of the handset under software control, such as the display, camera, audio, battery, etc. (the *handset peripherals*) and to operate hardware devices for peripheral communication capabilities such as USB, GPS, BT and WiFi, which may either be supported by separate chipsets or integrated functions of the main handset chipsets (the *communication peripherals*).

3.5.5 Mapping software architecture to chipset architecture

For low-cost handsets (as well as most feature phone designs up until the late 2000s), the software architecture indicated in Figure 3.3 executes on a chipset containing one microprocessor and one DSP. The DSP is responsible for much of the encoding and decoding of digital information to and from the radio transceiver. The microprocessor is responsible for managing the signaling of information between the handset and the network via a well-defined protocol stack which implements the 3G/4G wireless standards such as WCDMA, CDMA2000 or LTE, along with legacy 2G standards such as GSM and CDMA (IS95).

For such a low-cost phone, the management of peripherals such as the display, camera, input (touch or keypad) and audio is also managed from the same processor, with the option of companion processors to manage most of the peripheral functions such as GPS, Bluetooth, FM radio and WiFi.

By contrast, for all smartphones today, as well as high-end feature phones, there is now a clear divide between an application processor chipset, which executes the *application domain* software as well as the vast majority of the *device driver domain* software, and a modem chipset (microprocessor plus DSP), which is dedicated to executing the *wireless modem domain* software. With this architecture, the modem processor contains a lightweight real-time operating system, whereas the application processor contains a high-level operating system (HLOS) such as Android, iOS, Symbian, etc.

The wireless modem domain software is structured as a set of layers, which together form a protocol stack. A protocol stack is a well-defined concept in telecommunications and is a powerful way of supporting a complex set of interactions between two systems – in our case, a handset and a core mobile network. This critical concept of software layers is explained in more detail in Chapter 5, "Software design."

3.6 Manufacturer view

The fifth point of view we shall take is the manufacturer view. In this section, by manufacturer we mean the organizational entity responsible for building the physical handset from its component parts, along with the handset's standard accessories and packaging, into a form which can be shipped into the sales channel. The manufacturer traditionally has been an organizational entity owned by the handset brand – so a Nokia handset would be made in a Nokia factory and a Samsung handset would be made in a Samsung factory. However, in a significant number of cases, the manufacturer is a separate organization to whom the handset brand sub-contracts the manufacture of the handset.

3.6.1 Key components

At a high level, a manufacturer views the anatomy of a handset as comprising the following key parts:

- a set of electronic components, which must be mounted onto a printed circuit board (PCB);
- a set of mechanical components, including display, keypad, camera and caseworks, which must be assembled to construct the physical product;
- software, which must be programmed into the handset, to provide automatic test of the handset, as well as the particular customized version of the handset software required for the target market or customer (e.g. a network operator);
- a set of accessories, such as battery, power supply, hands-free headset, serial cable;
- a set of documentation, such as user manual and market-specific written materials;
- a set of packaging, including the final "sleeve," which may be market-specific.

3.6.2 Standard product versus customized product

In order to manufacture a modern consumer electronics product cost-effectively, it is essential to automate the complete production process almost entirely. Factories are very capital intensive to build and run, and, with margins often slim, profitability is achieved by keeping the factory highly utilized with a high volume of production. Pushing in the other direction, against this manufacturing need, is the market need to have many different products, as well as many different variants of each product, in order to provide high levels of differentiation and localization, potentially across the globe. This critical tension between standard product and differentiated product results in a number of fundamental handset design issues, which have to be understood and dealt with in the early stages of handset component design. Due to the very high volumes of mobile handset production, these issues are accentuated; the requirement to make a decision on whether to fit this component or that component creates a manufacturing inefficiency, which, if not contained, can become a significant cost when multiplied by the volume of product being created.

Much like other industries, such as the car industry, taking a platform approach also allows many common components to be shared between products. In turn, these common components can be manufactured in much higher volumes, which, due to economies of scale, lead to a downward trend in cost.

As a rule of thumb, the best way to achieve product customization is to do so in software rather than hardware, and in branding and packaging rather than mechanical design. Some examples follow.

For many years, Nokia had very successfully organized their portfolio of products into a small number of platforms or "series." So, for instance, all Series 40 phones use a black and white or color screen with the same mechanical characteristics and the same screen resolution. The keypad design is standardized. This in turn allows a more or less common user interface approach to be created just once in the software platform (thus gaining strong cost efficiencies in amortizing the cost of the software platform across many millions of devices). However, within this standardized approach, many different software customizations are possible – for instance, to include or exclude software for certain handset functions which require additional hardware support such as FM radio, Bluetooth and 3D graphics, to support different packs of supported languages for regional markets, and to use (or not use) certain hardware features of the underlying platform for market segmentation reasons (e.g. dual band, tri band or quad band). There are clearly limitations to this platform approach, and notably it was widely commented that Nokia's early resistance to responding to a trend towards "clamshell" form-factor phones was in part due to the large investment costs involved in creating a new clamshell platform, as none of the existing platforms – with all of their inbuilt efficiencies – were suitable. As a solution to dealing with this new platform requirement, Nokia chose to work with ODM partners to create a range of clamshell designs, with software adapted to the Nokia "look and feel."

In addition to altering the functions of the handset by changes to a common software platform, strong product variation is created through different industrial design approaches – allowing many different products to be created in a cost-effective R&D fashion, and, perhaps more critically, in a manner which allows very efficient manufacturing. Again, there is a downside to this approach. As we will see later in this book, the preferred approach to designing a handset is from the "outside in" – with brand DNA driving the "look and feel" of the product in terms of industrial design and user interface. If the software and hardware platforms are too constraining, then some hard limits will be placed on the level of product differentiation that can realistically be achieved, alternatively, costs rise dramatically in an attempt to modify a standard platform to meet a market need.

Regarding the PCB design, it is possible to plan to not fit certain expensive components if they are not required for a particular product variant. So, for example, the high-end variant of a product may contain capabilities to support Bluetooth, GPS, WiFi and FM radio. Mid- and low-end variants of the product may have reduced functionality, or less memory, in order to achieve a lower price point. Clever design can provide the means to not fit certain of these peripheral components, whilst preserving the overall integrity of the design (simply put: if you remove components, you need to ensure that the remainder of the handset will still work correctly).

3.7 Operator view

The final point of view we shall take is the operator view. To a network operator, the mobile handset is both the center and the edge of their customer service proposition.

The handset is at the center, because it is the one tangible, physical manifestation of the service relationship that the network operator has with their subscriber. The subscriber must make use of the handset to use any of the services of the network operator, be that voice calls, texting, e-mail, web browsing or social networking. On the other hand, given the complexities of mobile network infrastructure and the seeming "magic" of being able to connect a subscriber from one phone anywhere in the world to another subscriber in a completely different location, the mobile handset is often referred to as the "mobile terminal." As we saw earlier, in telecommunications, a terminal is a piece of equipment which exists at an "end-point" in a network – by definition, it exists at the edge of the network. The mobile terminal is the origination or destination point for a mobile telephone call which has passed through many switches and gateways in order to connect one end-point to another. Furthermore, handset products come and go with great rapidity, with users changing their handset every year or two, compared with an operator's investment in network infrastructure which must support millions of users and be sustainable over periods of decades, albeit with on-going renewal and upgrades to the network.

So, whilst recognizing the absolute importance of the handset to their service proposition, network operators have something of a love–hate relationship with the handset. The handsets must be purchased from handset brands, which in turn seek a strong brand affinity with their customers, often in conflict with the brand affinity which the network operator is also seeking to develop with the same customer. In many markets, the network operator will subsidize the cost of a mobile handset in order to make the entry price point for taking on board the operator's services much lower. This market dynamic means that it is strongly in the operator's interest to drive handset costs as low as feasible, as handset cost is a significant contributor to the cost of business, even before the subscriber has made a single phone call.

On the other hand, given that network operators have to make large investment decisions on infrastructure, often many years ahead of service launch or take-up, it is strongly in the network operator's interests to secure leading-edge handsets. Operators always need handsets that provide the means for subscribers to make use of an operator's next-generation capabilities, such as higher bandwidth data transfer, improved voice codecs, or advanced multimedia capabilities and browsers which provide an engaging online mobile Internet experience. These leading-edge handsets are complicated and expensive for the handset manufacturers to make, with volumes starting off low compared to more mainstream products. Many "in jokes" have existed over the years as operators have desperately sought next-generation handsets to work on their new networks. Examples include "God send mobiles" (GSM) and "late, tempting and elusive" (LTE).

3.7.1 Operator customization

In addition to the supply and demand market dynamic between handset manufacturers and operators, operators often see the need to add differentiating features to handsets sold on their networks, normally to seek to make the provision of operator-specific services more integrated into the overall handset user experience. In practice, handsets may well be standard at the air interface and network protocol level, but operators often place additional requirements on handset manufacturers to add further differentiation. Alternatively, operators may utilize standard handset capabilities (such as web browsing), but work with the handset vendor to ensure that these applications are adapted to interact with the network to deliver the best user experience and network performance.

Depending on the purchasing power of the operator, there are a number of points on the scale of customization which a handset vendor may agree to undertake on behalf of the operator, in return for the operator's business. At the more straightforward end, this might mean pre-programming phones with the entire network configuration settings required to access operator services such as SMS and an operator's portal. Alternatively, there may be links in menu structures or on the "home screen" of the phone which provide shortcuts for the handset user to access the operator portal. Moving along the scale, operators may have applications developed which reside on the handset and support their service proposition – an example here would be a device-resident application store for a large operator such as Vodafone or Verizon. The other end of this scale would be where many or most of the handset applications, and the complete handset user interface, have been developed specially for the network operator. This model of very deep operator involvement in handset design has been most common in Japan. Large operators or operator groups elsewhere, notably the Vodafone Group, have adopted a similar approach in order to own the whole end-user service experience.

In closing, we observe that it is a very significant challenge for both the handset manufacturer and for their key suppliers of software platforms, displays, etc. to be able to support the need for very customized handsets, whilst ensuring that they are able to achieve the price points the market requires through standardization and high-volume production of the same product. This critical tension is one of the most fascinating aspects of the design and supply of modern mobile handsets. The supply chain of component manufacturers and platform suppliers has to balance these issues so carefully to achieve the paradox of manufacturing the world's highest volume consumer electronics product by far, with probably the world's most differentiated product – the phone in your pocket.

3.8 Summary

In this chapter, we have examined the structure and organization of a mobile handset from a number of different perspectives: as a product teardown, as a hardware design, as a software design, and from the perspectives of a standardization body, a handset manufacturer and a network operator. We have discovered that although each perspective provides a self-consistent viewpoint to understand the structure of a mobile handset, no single perspective provides a comprehensive picture. We believe that to design successful technology components and mobile handset products, a thorough understanding of each of these perspectives is required in order to understand fully the market need.

In Chapters 4 and 5 we examine in more detail the core hardware and software components of a mobile handset, and the challenges of creating this core technology in such a way as to enable a great variety of different handset designs from many different organizations.

4 Hardware design

4.1 Introduction

In this chapter, we look at the major internal hardware components of a mobile handset and the key design issues which affect their design. We shall focus on core design aspects which are most particular to mobile handsets – radio, baseband cellular modem and mobile application processors. We shall look only briefly at the many – yet critical – peripheral components such as display, camera, audio, GPS, WiFi, sensors and so on. Our justification for taking this approach is that the core hardware and software components are the essence of what makes a mobile handset different to other consumer electronics products such as digital cameras, portable games consoles, SatNavs, set top boxes, etc. Many of these other modern consumer electronics devices also have displays, cameras, audio, memory, interfacing to other devices and connectivity options, so we shall focus on those aspects which most differentiate a mobile handset. In doing this, we are not, by any stretch of the imagination, under-estimating the importance which these other components play in creating a contemporary design. Indeed, these components are subject to the same core design constraints as the core handset components, and component manufacturers create peripheral products which are specifically suited to the design constraints of mobile devices.

4.2 Helicopter view

At a high level, Figure 4.1 illustrates the key functional hardware blocks which make up a modern handset such as a smartphone.

The *application domain* is primarily responsible for all of the interactions with the end user, including screen, camera, input, user interface,

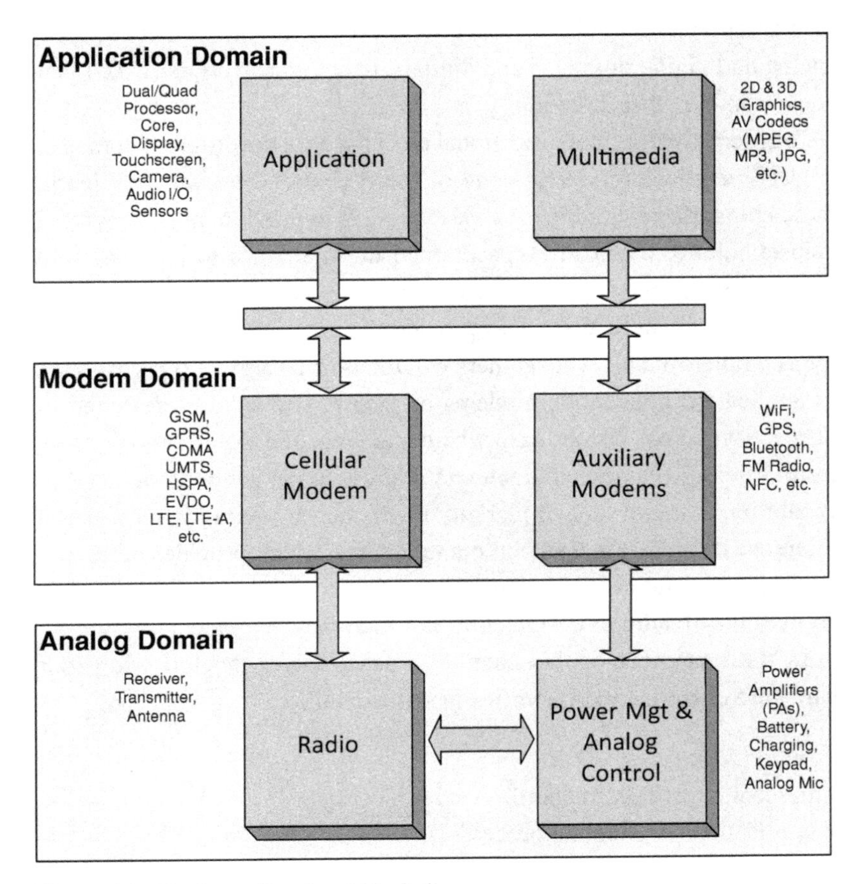

Figure 4.1. Hardware functional block diagram.

applications, menus, graphics, audio-visual and, if present, the underlying mobile operating system.

The *modem domain* is primarily responsible for all communication of data (voice, multimedia, Internet, etc.) and the control of that communication, between the handset and other parties. In the case of the cellular modem, this is the cellular network operated by a mobile network operator. In other examples, it may be connecting to a Bluetooth printer, a GPS satellite or a WiFi access point.

The *analog domain* is primarily responsible for all the interactions with analog hardware such as antenna, battery, power amplifier, keypads, as well as analog-to-digital converters (ADC) to convert between the analog and digital domains, and digital-to-analog converters (DAC) that convert in the other direction.

In order to realize these functional blocks within the different domains, chipset manufacturers create one or more chipsets. As we shall learn, these chips can be combined in various ways to provide more integrated chipset solutions, or can be partitioned in other ways to allow a more modular or flexible approach across a range of handset product designs. Some silicon manufacturers may provide all the chipset solutions required for a complete phone, whilst others will focus on an area of expertise such as application processors, modems or radios. The handset designer, in turn, must make decisions as to whether to "mix and match" components from different silicon manufacturers in order to get the optimal blend of capability, cost and time to market, or whether to take a platform-based reference design from a silicon manufacturer, which provides more of a ready-made solution and a much faster time to market, but necessarily reduces the breadth of design choices which may be made.

In the remainder of this chapter we shall cover the following four hardware component capabilities in some detail:

- radio (RF);
- baseband (cellular modem);
- application (application processor);
- multimedia (multimedia processor).

We shall cover in much briefer detail the following two functional blocks:

- auxiliary modems: GPS, WiFi, Bluetooth, IrDA, FM radio;
- analog: antenna, battery, keypad.

We shall briefly overview the following key peripheral components of a handset:

- user interface: display, touchscreen, camera, speaker, microphone, vibrator;
- multimedia output: stereo audio, HDMI;
- local connectivity: USB, NFC;

- sensors: gyroscope, accelerometer, compass, light;
- memory: external memory, flash, SRAM and EEPROM.

4.3 The radio spectrum

A mobile handset transmits information to and receives information from a cellular base station via the radio spectrum. Before we examine radio design issues, we need to ensure that we are clear on what the radio spectrum actually is, its characteristics, and how information can be transferred via radio waves.

4.3.1 Spectrum

By national and international agreements, different ranges of frequencies of the radio spectrum (known as "bands") are allocated for different purposes. For example, analog and digital broadcast television, cellular telephony, emergency services and the military are each allocated "bands" of the spectrum by the government for specific uses. Figure 4.2 provides an illustration of the electromagnetic spectrum, with particular reference to the radio spectrum. Standards also exist, which all products must conform to, which are designed to reduce or eliminate the spurious transmission of unwanted signals outside the allocated band for a product, and guarantee a minimum level of performance. Hence televisions must not emit (beyond permitted levels) electromagnetic radiation which could interfere with other equipment such as mobile phones, and vice versa.

4.3.2 Radio transmission

Just like a vehicle's tires are "where the rubber hits the road," and the vehicle physically connects with the road transportation network, so it is with a mobile handset that the radio component is where the handset interacts with the radio spectrum, and thus connects to the mobile telecommunications network. The radio is usually unseen, certainly unappreciated and definitely a mystery to the uninitiated. However, without

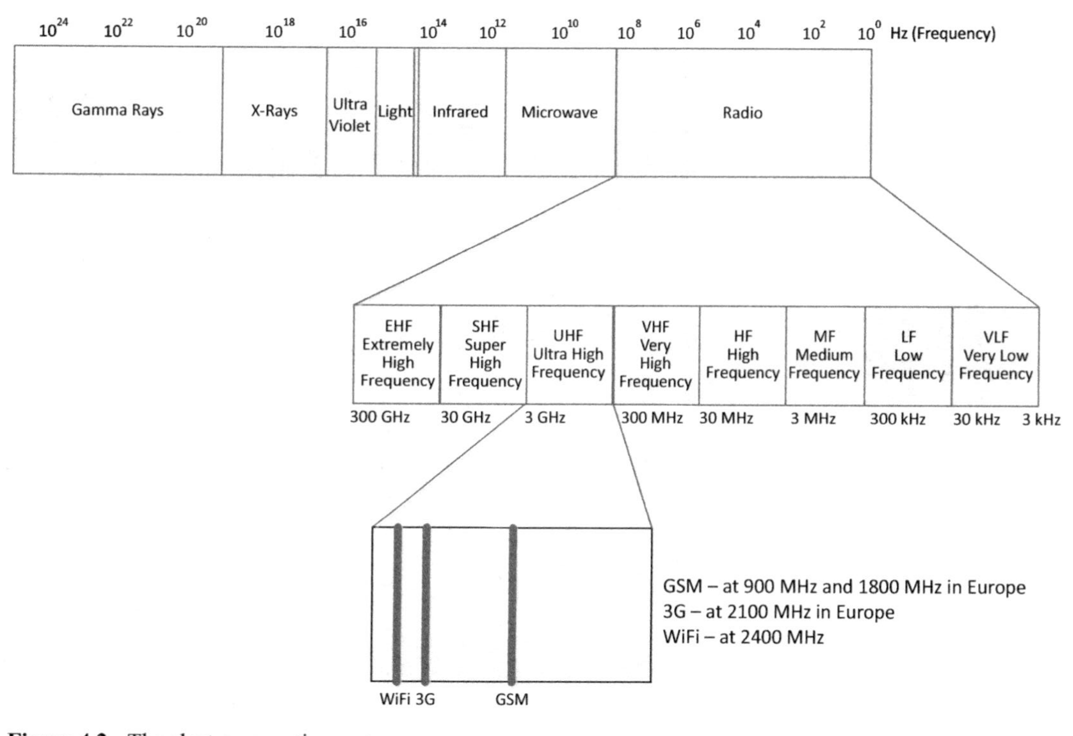

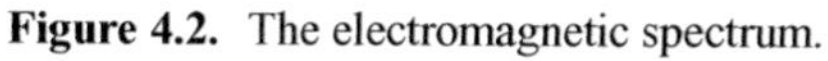

Figure 4.2. The electromagnetic spectrum.

the development of radio and other electromagnetic transmission and reception technology over the last 100 years, the modern world would be very different. We would not have more than six billion mobile phone connections in the world with over 1.5 billion new handsets sold every year. There would also be no television, no favorite radio station, no cockpit to ground station communication in aircraft, no satellites, no stunning pictures from the Hubble space telescope . . . and no WiFi in the coffee shop.

Radio waves are electromagnetic waves, which exhibit a broadly common set of properties with wave frequencies in the range 3 kHz to 300 GHz. Radio frequencies used for cellular networks operate in the UHF (ultra-high frequency) band, which lies between 300 MHz and 3 GHz. The basic, unmodulated, radio wave appears to measuring equipment as a received signal which oscillates between an amplitude maximum and an amplitude minimum at a fixed frequency. So a radio wave at 900 MHz oscillates 900 million times in the space of one second. Such a fixed frequency wave is called a *carrier signal*.

The distance that the wave can be transmitted is a function of the transmit power (i.e., the energy that is input to propagate the wave) and the propagation characteristics of the medium in which the waves travel (e.g. through a vacuum, through air or through the walls of your house).

Information can be encoded into a signal by modifying over time one or more of its fundamental characteristics:

- frequency (how often the wave oscillates – frequency modulation, or FM);
- amplitude (the amount of energy used to create the height or "peak" of a wave – amplitude modulation, or AM);
- phase (where on the timeline the wave pattern repeats – phase modulation, or PM).

The encoded signal is created at a low, or *baseband*, frequency, using an appropriate modulation scheme such as frequency modulation. However, in order to transmit the encoded signal, it must be translated into a high frequency within the radio spectrum in the range of megahertz to gigahertz. This is achieved by *mixing* the two signals – the modulating

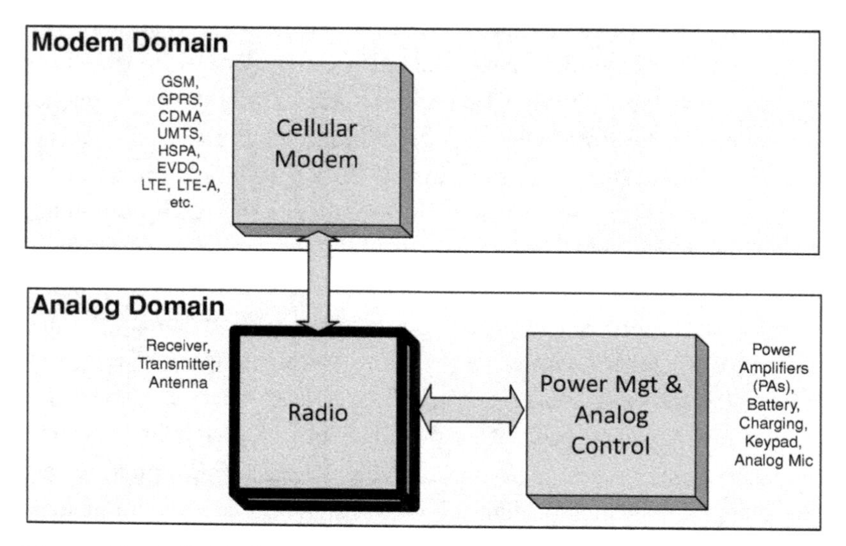

Figure 4.3. Radio block locator diagram.

signal at baseband, which is rich in encoding information, and a simple sinusoidal wave at a significantly higher radio frequency. The result of this mixing is known as *modulation*, and the resultant signal is known as the *passband* signal. Modulating onto a sine wave makes it possible to keep the frequency content of the resultant signal as close as possible to the carrier frequency of the passband, making efficient use of the available spectrum.

4.4 Radio chipset design

In this section we look at the major factors which influence the design of the RF or radio chipset of a mobile handset. Figure 4.3 indicates the interface points between the radio hardware component and other parts of the handset hardware. At the beginning of each remaining section of this chapter, we include a diagram in a similar format that shows the component to be discussed and its main interface points to other components. Each of these diagrams is a subset of the overall hardware block locator shown in Figure 4.1.

4.4.1 Radio transmitter

Given the nature of radio transmission, we can outline the key functions of a radio transmitter as follows:

- to take as input an encoded signal (typically from a digital baseband component);
- to generate a carrier frequency at the required output frequency;
- to combine the encoded signal and the carrier signal into a high-frequency modulated signal;
- to transmit the resultant frequency at sufficient power to propagate the signal a required distance, without unrecoverable loss of the information content of the signal.

This simplified model of a radio transmit function assumes that the signal is transmitted at a constant power, for a continuous period of time, at a constant carrier frequency. In reality, however, cellular radio systems vary (*modulate*) each of these parameters in order to optimize radio performance. The fundamental design trade-off is between an increased transmit power, which improves the likelihood of the signal being transmitted without loss of information, versus a reduced transmit power, which minimizes both the interference caused to adjacent frequencies and the current drawn on the battery of the device.

Note that GSM originally chose a modulation scheme known as GMSK (Gaussian minimum shift keying), because it requires a constant transmit power. This enabled maximum efficiency in the transmitter power amplifier, which fitted the technology of the day. Later, a modulation scheme with an amplitude component was defined as part of EDGE (enhanced data rate for GSM evolution – a 2.5G standard for packet data). This gave higher data rates, but required a major re-design of the transceiver.

A constant transmit power only makes sense if the "end-point" which is to receive the signal is a constant distance away from the transmitter and the propagation characteristics between the transmitter and the receiver are constant. In a mobile communications system, the distance between a mobile handset and a base station will change as the user of the mobile handset moves from place to place. Further, the radio propagation characteristics vary over time, due to changes in distance between transmitter and receiver, due to interference from other (reflected) signals and due to changes in the surrounding physical topography (buildings, hills, etc.) and even atmospheric conditions. The cast-iron method to deal with these changes in propagation capability would be to transmit at the highest power achievable all the time. However, this would have numerous negative effects, beyond the obvious wastage of energy and therefore reduction in mobile handset battery life. Transmitting at a power higher than necessary can result in RF interference being caused in other products operating on adjacent channels, and consequently permissible transmit power is tightly defined and regulated in mobile communication standards.

Transmitting on a continuous basis is only energy efficient if there is a continuous stream of information to transmit. In practice, the nature of much communication is "bursty" – with periods of intense information transfer and other, longer, periods with no information to transfer. Hence, radio transmitters are designed to be able to be switched on and off rapidly, so that power is only consumed when absolutely necessary.

Transmitting at a constant frequency ensures that transmitter and receiver can be tuned to the same frequency once for the duration of the transmission. This is the basis on which traditional broadcast systems operate – such as broadcast radio and television. However, there is a key disadvantage to this approach. If the frequency chosen for transmissions happens to be suffering from bad interference from other signals, then there is a risk of data being lost in transmission. For radio broadcast, this means that if reception is poor on one frequency, then the radio user can re-tune their radio to an adjacent frequency, which may have better propagation and, therefore, reception capabilities. However, in a mobile communications system, where transmitters are operating at a much lower power than broadcast systems, and with a high degree of mobility of handsets, the propagation characteristics at a particular frequency will vary considerably over time in a way that cannot be predicted. For these reasons, digital radio standards for mobile provide a range of mechanisms for the transmit (and receive) frequencies to be changed over time, in order to smooth out the propagation advantages and disadvantages of different frequencies. This deliberate changing of frequencies during transmission is known as *frequency hopping*.

In the GSM standard, frequency hopping is employed according to a pre-agreed *hopping sequence* with the receiving end. In the case of the Bluetooth standard, continuous frequency hopping occurs to minimize interactions with other devices which may be transmitting in the same ISM band. The ISM band is unlicensed and thus may be used by a range of different low-power devices, be they Bluetooth or not. A related approach is taken in CDMA-based standards (including 3G), where information is transmitted (or spread) across a broad range of frequencies as rapidly as possible, according to a sequence created by a code generator. This is the origin of the term "spread spectrum" – the information to be encoded is spread across a large number of frequencies. With the ability to vary the frequency used to transmit, as well as the period of time and power at which transmission occurs, we can begin to view the radio spectrum available to a mobile communications system as a shared physical resource, which can be accessed in a number of different ways. The available set of contiguous frequencies of this physical resource is known as the "RF band." In many modulation schemes this RF band is then divided up to be used by a number of RF carriers, each with its own spectral bandwidth. By contrast, in CDMA-based systems, the whole of the available RF band is used to spread out the information content rapidly.

4.4.1.1 Physical and logical channels

Now we need to introduce two different concepts for how we describe a channel. The first is a particular radio frequency which can be used to transfer information, known as a *physical* (or *RF*) *channel*. A physical channel represents a particular fixed frequency, within the RF spectrum, which has been allocated for use for the transfer of information between systems which may transmit and receive signals (called a *transceiver* – a combination of the words *transmitter* and *receiver*).

A second concept is that of a *logical channel* of communication between a transmitting device and a receiving device. A logical channel represents the flow of information between one transceiver and another, irrespective of how the information is physically transferred. In the most simple form, one logical channel can be mapped to one physical channel, as was used in the early 1G systems described in Chapter 1. From the 2G standards onwards, logical channels have been typically mapped to different physical channels in sophisticated *channel access* methods. The key reasons for adding complexity to channel access are to make the most efficient use possible of the available radio resource (to increase the number of users who may use a particular cell) and to maximize the information throughput, whilst reducing interference – both from other users and from background RF noise.

4.4.1.2 Channel access methods

With this concept of a logical channel, we now briefly describe the main types of method used to *access* the spectrum resource in order to transmit data through the logical channel, via physical channels, to the receiver.

In descriptive terms, each of these access methods may be summarized as in Table 4.1.

4.4.2 Radio receiver

At the most fundamental level, the radio receiver must reconstruct the information at the receiving end that was previously transmitted by a transmitter unit. The transmitted signal is propagated as a radio wave. As propagation occurs over a distance, the power of the signal in a vacuum reduces according to an inverse square law. In practice, the propagation typically occurs not through a vacuum but through a range of substances (air, brick, glass, etc.), each of which will result in further loss of energy of the radio wave. The combination of these energy losses is known as *path loss*. Almost inevitably, there will be other radio waves present at the same frequency as our radio wave – these could result from other transmitter systems, from electromagnetic interference from motors and other equipment, and not least from background radiation, either originating from the Earth, or (quite likely) from outer space (so-called "cosmic rays"). These "unwanted signals" will coexist with the "wanted signal," and will lead to varying degrees of *noise* being included in the *signal* received by the antenna. The term *signal to noise ratio* is used to represent the amount of noise present with respect to the strength (energy) of the wanted signal. Having received a signal at the antenna, it is a prime role for the receiver to remove unwanted signals and to

Table 4.1. *Channel access methods*

Mnemonic	Description	Usage
FDMA	Frequency division multiple access – shares the spectrum resource by allocating each user to use a different physical channel (RF frequency) to everybody else in order to transmit or receive information	The 1G standards ETACS and AMPS use FDMA The 2G standard GSM uses both FDMA and TDMA FDMA is used to separate the frequencies between different adjacent cells
TDMA	Time division multiple access – shares the spectrum resource by dividing it up into a set of physical channels. Each physical channel is separated over time into different periods or timeslots. Users communicate information in an allocated timeslot, resulting in a number of users sharing the channel resource, but with only one user transmitting or receiving information at any one time on a particular channel	The 2G standard GSM uses both FDMA and TDMA TDMA is used to share the channel resource between different users

(*cont.*)

Table 4.1. (*Cont.*)

Mnemonic	Description	Usage
CDMA	Code division multiple access – shares the spectrum resource by allowing each user to utilize all the available channels. Use of the channels is shared by spreading each communication over multiple channels using a pseudo-random coding sequence which is designed to minimize interference to other users, allowing multiple users to communicate information at the same time over a wide frequency range	The 2G standard IS95 and 3G standards WCDMA and CDMA2000 use CDMA
OFDMA	Orthogonal frequency division multiple access – a large number of separate (orthogonal), small and closely spaced physical channels are used to communicate information at a low data rate. Each user is allocated a number of these physical channels to communicate information, which, when taken together, provide a much higher data rate	The 4G standards LTE and WiMax use OFDMA

reconstruct the wanted signal in as close a form as possible to the signal which was first transmitted by the transmitter unit.

Filters are used to remove information from the received signal. For example, a band pass filter allows all frequency components within the required bandwidth of the signal to pass through, whilst removing frequencies which are higher or lower than the wanted signal. High or low frequencies within the band of the wanted signal can also be removed with high and low pass filters, respectively.

A key issue remains, which is how to reconstruct a corrupted signal. The signal may become altered through the addition of unwanted interfering signals and background noise, and the receiver cannot know what the transmitter was seeking to communicate ahead of time. Typically the nature of signal corruption is that it occurs for a short period of time (perhaps just a few bits of data), resulting in a section of the transmitted signal becoming corrupted, with uncorrupted data either side of it. Given the nature of this signal corruption, an approach known as *interleaving* is used. This involves "chopping up" the digital signal to be transmitted into a number of equal sized blocks. Like a Rubik's cube where the colored squares have been mixed up, the order of the blocks is then rearranged, spreading out any particular section of signal over a longer period of transmission time. The size of the blocks and the rules for rearranging the blocks are designed so as to reduce the possible *sequential* data loss through signal corruption which could occur when the interleaved sequence is transmitted and reconstructed (*de-interleaved*) at the receiving end. With de-interleaving, the severity of signal corruption for a section of data can be reduced, but not eliminated.

A further approach is employed known as forward error correction (FEC). This technique involves building extra information (known as redundancy) into the signal, which can help to provide "clues" that allow the original signal at the receiving end to be reconstructed, even if it has been corrupted during transmission. The maximum number of missing bits that can be corrected is dependent upon the design of the FEC algorithm. There is a design trade-off between the number of bits which can be recovered versus the extra capacity needed on the channel to encode redundant bits.

Forward error correction techniques allow errors in the original signal to be detected and corrected. If the data corruption is beyond certain limits, then the approach is to identify the section ("block") of data which is corrupted beyond repair, and for the receiving end to signal to the transmitting end that the data has become corrupted and should therefore be transmitted to the receiver once again.

There are two common approaches to forward error correction; one is known as block encoding and the other as convolutional encoding. Block encoding involves dividing up the signal into blocks of bits and protecting each of those blocks with redundant bits which can be used for error correction. Convolutional coding works on bit or symbol streams of arbitrary length and adds redundancy (additional bits) to the data stream, using a section of the original bit stream to create the codeword to use to encode the data. The resultant bit stream can be decoded at the other end, where the decoder can withstand the corruption of a number of the data bits whilst still being able to recover the original data stream. The almost universally used algorithm for recovering the original bit stream is known as the Viterbi algorithm. Although very computationally intensive, the Viterbi algorithm lends itself to the algorithm being run in parallel (computing a number of steps at the same time, without interdependency between those steps), and is very well suited to implementation in silicon.

The most advanced form of coding in use today is known as turbo coding, which is the name given to a set of high performing forward error correction techniques that approach the theoretical maximum (known as the Shannon limit) for the coding rate possible for a given noise level in the communication path. It was long thought impossible to develop coding schemes which could get anywhere close to the Shannon limit. However, since its development in the early 1990s, turbo coding is now extensively used in 3G and 4G systems, as well as satellite systems and modern space exploration.

4.4.3 Multi-band and multi-mode

The terms *multi-band* and *multi-mode* are in common use, and it is important to be clear on the distinction. A radio which is multi-band functions across two or more frequency bands (or sections) of the radio spectrum, using the same radio standard. The classic example here is GSM quad band, where a radio is able to transmit and receive GSM signals in the 850 MHz, 900 MHz, 1800 MHz and 1900 MHz bands.

A radio which is multi-mode functions across two or more different radio standards. Examples here would include GSM and 3G, or 3G and WiFi. Because spectrum allocation ensures that alternative standards do not coexist in the same space, a multi-mode radio also works across different bands, in order to support the different standards.

4.4.4 Envelope tracking

The radio subsystem of a mobile handset is becoming ever more complex, due to the increase in the number of radio standards requiring support – ranging across the whole breadth from 2G to 4G. In addition, there are an ever increasing number of frequency bands to support. Together, this is leading to a significant increase in RF system complexity, and, with higher data rates and "always on" connectivity, battery life has again become a major design challenge. The standard approach to addressing the increasing number of bands is to use more power amplifiers (PAs), each supporting a narrow band, in order to optimize the peak power efficiency for that band. However, this can lead to a theoretical eight PAs being required in some designs, leading to other design challenges such as RF switching times, handset form-factor and increased battery drain due to large increases in heat dissipation.

The traditional RF design approach is to design a PA with a fixed power supply voltage. With this design approach, the PA efficiency reduces as the ratio between the peak transmitted power and the average transmitted power increases – this is known as the *crest factor*. For handset designs, a new approach to this problem is a technique known as *envelope tracking*. Envelope tracking is a design approach applied to the PAs which constantly adjusts the power supply voltage applied to the PA to ensure that it is always operating at its peak efficiency, even as power output requirements change dynamically. Although envelope tracking itself is not new, having been around since the 1930s, advances in design techniques have now made it a viable approach for mobile handsets. It is claimed that 4G terminals without envelope tracking will require 75% more power than existing 3G terminals.

4.4.5 Antenna design

An antenna is a transducer which converts electromagnetic waves into electrical currents, and vice versa, according to the physical laws of electromagnetism. For transmission, an alternating electrical signal applied to the antenna results in the radiation or emission of electromagnetic waves. For reception, electromagnetic waves reaching the antenna induce an alternating current in an electric circuit connected to the antenna. Critically, the frequency of electromagnetic radiation which is transmitted, or may be received, is determined by the physical characteristics of the antenna, including its spatial dimensions, construction and material density.

Nearly all modern mobile devices contain an internal antenna within the casework of the device itself. This contrasts with earlier antenna designs, where the antenna was typically a "stub" which protruded from the top of the handset. There is an obvious ergonomic advantage in losing a protrusion from the handset, more so as handset sizes reduced significantly in the 1990s, leading to mobile devices being carried easily in a pocket or a handbag.

In modern antenna design, there are the following five key design parameters to consider and balance.

- The transmitter power. Given the battery and size constraints of a handheld device, how effectively the transmit signal can be propagated is a key design parameter. By contrast, the base station does not have these same design constraints to work to.
- The receiver sensitivity (how weak a signal can be perceived, for a given noise level).
- The frequency band across which the radio has to be able to transmit and receive RF.
- The 3D cavity available into which the antenna must fit, dictated by the industrial and mechanical design of the handset.

- The requirement for transmission to direct the majority of radiation away from the handset user. This requirement is normally specified as a measure of the rate at which energy is absorbed by the body when exposed to radio waves and is known as the specific absorption rate (SAR).

Poor consideration of antenna cavity design can compromise the ability to realize an antenna design with good sensitivity. Sensitivity is a differentiating capability of a particular handset design as it is influenced strongly by the industrial design and the supported bands, both of which are specific handset design issues. The more bands that need to be supported, the more the antenna design has to compromise sensitivity for flexibility.

An exciting development in antenna research and design is MIMO, which stands for Multiple Input and Multiple Output. With MIMO, multiple antennas are used for both transmission and reception, with the objective of improving overall spectral efficiency (measured as the number of bits per second per hertz of bandwidth). The antennas must be separated from each other, to ensure that each antenna transmits or receives different signals, as a result of the signal departing or arriving via different paths (through, for instance, reflections). Simplistically, the information to be transmitted can be split between different antennae which can radiate the signal in different ways. At the receiving end, antennas can receive the different signals, arriving via different paths, and reconstruct the original signal. Methods are available for coding data into the different streams in order either to increase the effective data rate or improve the ability to recover the signal at the receiving end (or a combination of both). In order to achieve the higher data rates specified in standards such as LTE, it is now necessary to design handset products with this multiple antenna approach. In handset design, it is necessary to provide physical separation between antennas supporting different transmission standards. For example, it is common for Bluetooth to be active (to support a wireless headset) whilst making a cellular call. However, the Bluetooth receiver would easily become overloaded with the cellular transmit signal (even if at a different RF frequency)

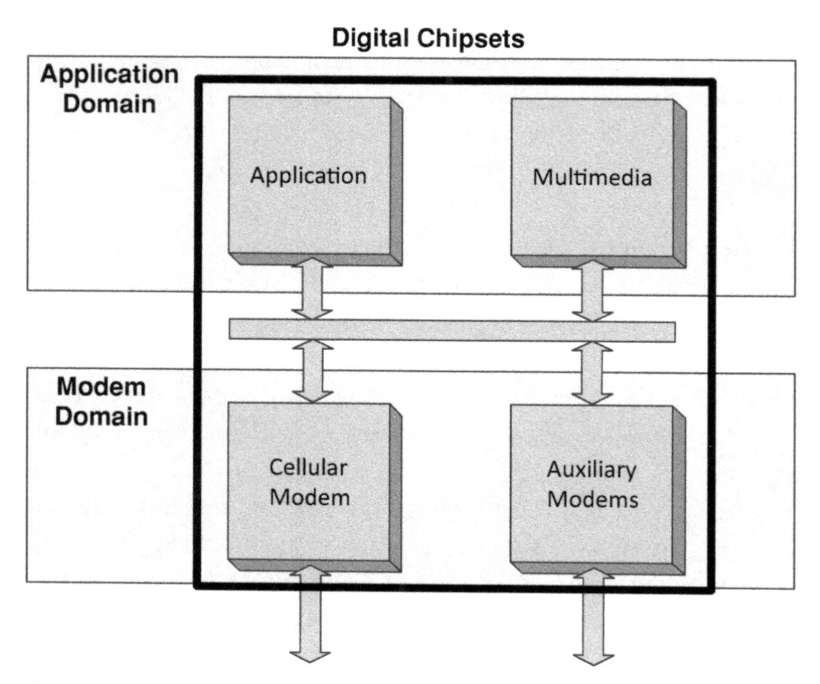

Figure 4.4. Digital chipsets block locator diagram.

unless there is suitable physical separation (and shielding) of the different antennas.

4.5 Digital chipset design

In our system block diagram there are up to four digital chipset blocks – application, multimedia, cellular modem and auxiliary modems (see Figure 4.4). These blocks are functional areas; each semiconductor manufacturer will make their own judgments on whether to design four different chipsets, one for each block, or whether to combine some or all of these functional blocks into a common chipset.

4.5.1 Key design parameters

In this section we present an overview of the digital chipsets that take a central role in providing the features and capabilities of a handset design.

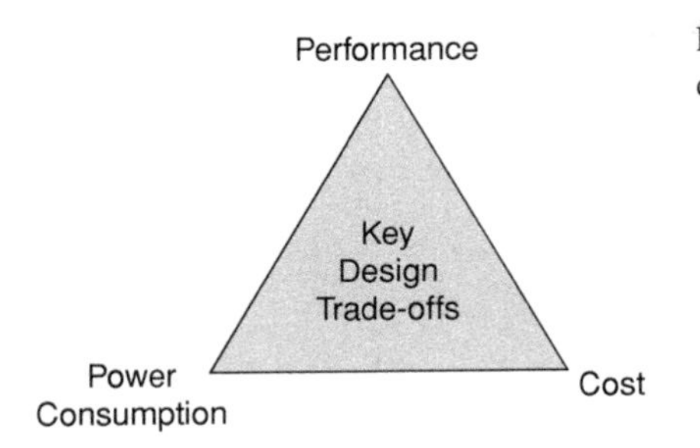

Figure 4.5. Fundamental chipset design trade-offs.

Key to success in mobile chipset design is achieving the optimal trade-off between power consumption, performance and cost (see Figure 4.5) – with good market timing in order to achieve the business objectives of volume, margin and competitive positioning.

Integrating more transistors onto the same surface area of silicon die allows more functionality for the same cost. Mounting a chip in smaller packaging allows for smaller or thinner handset designs.

For mobile devices, unlike PCs and other mains-powered devices, power consumption is the key design issue which dominates many decisions. Trends in chipset design continue to reflect increasingly innovative approaches to reducing power consumption further. A continued focus on reducing power consumption is essential in order to mitigate the great increase in processing requirements to support functions such as high-performance graphics and multimedia, and to provide support for an increasing number of wireless air interfaces.

4.5.1.1 Semiconductors

Advances in semiconductor technology are normally represented by Moore's Law, which states that the number of components per chip doubles roughly every two years. This increase in the levels of integration has provided a remarkable pace of improvement in all of the key semiconductor design parameters of cost, speed, power, size and functionality, which in turn has continued to fuel the growth and success of all aspects of the modern digital world, including mobile communications. All of the key design improvements have resulted principally from the industry's ability to decrease dramatically the physical size required to represent the basic building blocks of transistors and gates used in integrated circuits.

Table 4.2. *Silicon geometry roadmap*

Process geometry (nm)	Year of first use
130	2000
90	2002
65	2006
45	2008
32	2010
22	2011
16	approx. 2013

Improvements in semiconductor lithographic technology permit improvements in the *semiconductor manufacturing process* (often referred to as "process" or "process geometry") used to manufacture the chips. Process geometry normally refers to the half-pitch (i.e., half the distance) between identical features in a DRAM memory array. For example, the 130 nm (0.13 μm) process was first introduced in 2000. Well-known PC chipsets from Intel, such as the Pentium III and Pentium 4, were designed using this process. Technology improvements permit reductions in the process geometry of approximately 70% every two to three years. The semiconductor industry roadmap is guided by the ITRS (International Technology Roadmap for Semiconductors) supported by the semiconductor industry. The approximate roadmap for further improvements in process geometry is given in Table 4.2.

During 2011, most mobile chipset designs were based on a 45 nm process. The latest designs supporting LTE require a 32 nm or 28 nm process in order to achieve the desired chip density to fulfill the requirements of LTE. New challenges exist, however. Below 90 nm, leakage of current (through quantum tunneling effects) becomes a dominant chipset design issue. There is consequently considerable focus on strategies to reduce leakage. One strategy is to partition the chipset design across different voltage domains by running different sections of the chipset at higher or lower voltages. Those sections running at a lower voltage will be clocked at a lower frequency – with less power in the section, there is proportionally less power to "lose" when the chipset is idle. Another technique is to switch off completely those sections of the chipset which are not required at certain times – for instance, turning off the 3G/LTE capabilities when the handset is in a "fallback" GSM only area of coverage.

4.5.1.2 System configurations

Based on the particular design constraints that apply to the handset market segments for which the chipset is targeted, there could be one, two or more digital chipset packages present on a handset PCB. Within a particular chipset package, there could be two, three or more processing *cores* and memory units present on one silicon die. This is known as system on chip (SoC). Alternatively, two or more chipsets from different silicon dies can be combined into one physical package to reduce PCB size and thickness. This is known as system in package (SIP). Each processing core consists of a processing unit, typically a microcontroller or DSP (see the box in Section 4.6.3 for an overview of the difference between the two) or fully discrete hardware logic, for example to provide a standardized encoding or decoding function. In addition, each core may also contain memory blocks, timing and peripheral control logic.

Silicon manufacturers may employ a number of these different approaches across a portfolio of chipset offerings in order to achieve particular design trade-offs for each handset segment.

Key design parameters which affect the decision on the number of chipsets and cores and the type of packaging approach include cost, time to market, performance and the degree of flexibility required in order to create multiple handset designs. In the following sections we look in more detail at the two most common types of digital chipset – baseband and application – and then examine a number of design issues affecting the design trade-offs made for different target markets.

4.5.1.3 Power consumption and emerging multi-core designs

In 2011, smartphones first entered the market powered by multi-core CPU designs – initially dual-core, with the expectation of quad-core devices to follow. From a chipset design perspective, the major motivation to move from a single-core to a multi-core design approach is to improve performance further, without a resultant increase in power consumption. Traditional improvements in chipset performance are achieved fundamentally by migration to smaller process geometries (resulting in a faster time to switch a transistor, as the signal has less distance to travel in the same unit of time) and increasing the clock frequency (resulting in more transistor switches for the same unit of time compared to a lower frequency). Increasing the clock frequency also increases power consumption, as more energy is used in the same period of time, with an increasing amount of that energy being lost in the form of heat dissipation. Sophisticated power management design techniques are used to reduce the clock frequency wherever possible (for example, when a handset is in standby or "idle mode") in order to reduce power usage.

In addition, techniques such as the use of faster and larger memory caches allow the processor to be more fully utilized, instead of losing clock cycles waiting for data to be loaded from external memory, which can take a number of clock cycles to achieve.

As increasing the frequency also increases power consumption (according to a mathematical power law), beyond a frequency of about 1 GHz new methods are required to achieve further improvements in performance without significantly compromising power consumption.

By introducing a multi-core approach, it becomes possible to run two (or more) functions at the same time for a particular clock speed. So, in theory at least, it should be possible to double the performance of a chipset for the same power consumption by moving from a single- to a dual-core architecture. In practice, there are constraints imposed by the nature of the tasks which can and cannot be run in parallel, and the software overhead to manage the sharing out of tasks, which reduce this theoretical performance improvement. Nonetheless, in a contemporary mobile phone, there are numerous opportunities to introduce parallelism. Simple examples include web browsing whilst decoding high-quality video and audio, whilst updating real-time map data based on new data from a GPS receiver. End-user benefits are clear in terms of faster webpage loading, smoother multimedia playback, faster task switching and an overall more responsive user interface. Reminiscent of the PC industry, but new for the mobile handset industry, chip design facts such as processor speed and the number of cores supported are now being marketed by handset companies to demonstrate that new products are able to deliver improved performance to the end user.

Additional software support is required for multi-core designs – normally in the OS – to allocate tasks to different cores to achieve maximum concurrency. For software developers, there is an added responsibility to create code which is multi-threaded, which means that two or more instances of the same code could be run at the same time. Examples here include multimedia decoding of information on a media-rich webpage which may require multiple data decodes of graphical or script-based information.

4.5.2 CPU architectures

Early CPU architectures for baseband processors were focused on providing good general purpose performance for the lowest possible power consumption and the ability to control a growing number of peripheral components. With the rise of the feature phone device class in the early 2000s, focus moved to accelerating application performance in areas such as Java and camera acceleration, whilst developers concentrated on realizing multimedia codecs in software, highly optimized for the underlying instruction set of the CPU. With the move to higher modem speeds and more advanced smartphone capabilities, CPU architectures have developed in different directions (to reflect the rather different needs of an application processor and a baseband processor).

Modern CPU architectures for application processors provide considerable multimedia acceleration capability at an instruction set level. This means that it is still possible to write software in a high-level language such as C/C++ or Java, but these program instructions are converted into specialized CPU multimedia instructions which can execute in a single "heartbeat" (cycle) of the CPU, rather than in a number of cycles.

Modern CPU architectures for baseband processors are focused on improving protocol stack performance above layer 1. With each generation of mobile broadband technology, the data throughput has dramatically increased, and this now means that, once the data has been successfully decoded by the layer 1 subsystem (consisting of both DSP and CPU aspects), there is a very large amount of data to move around the system. In particular, this means that the protocol stack is now becoming a bottle-neck for processing. In response to this, CPU manufacturers have been developing a number of techniques to improve performance. One such area is cache performance. A cache is a reserved area of computer memory which is fast to access – much faster than storing data in other forms of memory for longer term storage. Another area is branch prediction.

What is branch prediction?

A very common capability of a CPU is the ability to perform different actions based on the results of a conditional test. For example, *"If X is greater than 10, then do instruction A, otherwise do instruction B."* Because there is a test to be made, it is not possible to be certain ahead of time whether instruction A or instruction B needs to be fetched from memory and decoded in order to work out what to do next. Branch prediction allows the most common circumstance to be predicted, allowing the CPU to have the next instruction ready as soon as the previous instruction has completed. Clearly this means that some of the time the prediction will be correct and some of the time it will be wrong. The trick is to identify areas of code where there are many instances of a test having a particular answer and fewer cases of the test having the other answer. A simple example is as follows: "Store the number 100 in variable X. Keep taking 1 away from the number stored in X until the value stored in X is 0." In this case, the CPU needs to take 1 away from a number and then check to see if the answer is 0. If it is not 0, then the operation is repeated. In this example, it is plain to see that 99 times out of 100 the answer to the question "Is the value in X equal to 0?" will be "No." Hence, branch prediction allows the CPU to "know what to do already" in the case where the value of X does not equal 0, and thus the CPU can prepare the next operation much more quickly.

4.5.3 Architectures for ultra-low-cost handsets

Although new features are being added to handsets at a phenomenal rate of change, there remains a significant opportunity to meet a need for more basic handsets, typically for emerging economies such as India, China, parts of Southeast Asia and Latin America. This market is large (over 300 million handsets a year by 2015 according to ABI Research), and large volumes drive the cost point downwards very strongly.

This section has focused on the emergence of two discrete types of processor: a baseband processor and an applications processor. However, for markets where the feature set is stable (enabling higher levels of hardware integration with less software flexibility), the requirements are not too high (allowing optimization for cost) and the volumes are considerable (ensuring economies of scale), there is a different set of design parameters which drive a highly integrated chipset solution. Perhaps the more obvious step is to have a single CPU architecture, with a processor capable of handling the demands of both modem and application functions. A further step is to integrate the digital chipset and the radio chipset into a single design. To reiterate, this high level of integration only makes commercial sense when the volumes are high and the requirements are firm, as otherwise the evolution paths for each of the radio, baseband and application domains run at different speeds to each other. Very high volume chipsets for low-cost markets use older processor technology – perhaps two generations behind the state of the art.

4.5.4 Packaging

Approximately 50% of the cost of a chip is related to its manufacture. This means that the physical size of the silicon, as well as the number of pins (pin count) required, are major design considerations. A chip design can be considered to be *pin bound* if the ultimate size of the chip package is dictated by the number of pins required, or *silicon bound* if the ultimate size of the chip package is dictated by the size of silicon required.

There are various packaging technologies available in the electronics industry, with older packaging approaches still in use where appropriate. Home electronic hobbyists may still be familiar with the dual in-line package (DIP), where there is a row of leaded pins down both of the long sides of a chip package. This form of packaging was superseded in the 1980s with advances in highly integrated circuits, which required more pins to connect to the rest of the product circuitry. As a result, pin grid array (PGA) packaging was invented, which allowed a large number of pins to be present at the base of the package, for the same surface area of package. A refinement to PGA was then ball grid array (BGA), whereby the pins are replaced by very small balls of solder. For modern chipsets, the focus is on reducing both the size of the solder balls and the spacing between the balls – both approaches contribute to a reduction in the package size. Smaller packages are cheaper to manufacture, but also provide improved miniaturization of the final handset design.

Various methods of combining chips in a single package are also common. When multiple dies are placed together in a single package, this is called system in package (SIP). When multiple dies are combined on a small substrate, this is called multi-chip module (MCM).

SIP and MCM packaging have had success where the dies come from the same manufacturer – for example, for memory chips which combine flash and RAM in a multi-chip module. However, such packaging has struggled to take off for baseband and application chipsets, largely because of the supply chain issues of sourcing different dies, on different schedules, from more than one manufacturer.

A further packaging approach is *flipchip* – this allows two chips to be mated together, with very low cost of interconnect, and holds promise for combining baseband and memory or baseband and RF chipsets.

4.5.5 Software-defined radio

Modern baseband processors need to support a number of different modems. There are a number of variants of 2G, 3G, HSPA and now LTE to support. Although there is some commonality between standards, there are, nonetheless, always more functions to support in each generation of chipset. As a consequence, the physical die size of the chip created is becoming a key design issue. The larger the die size, the more expensive the chipset is to manufacture. With analog (RF) chip design, moving to smaller process geometries does not provide the same performance improvements that are seen in digital chips – in fact, noise from *intermediate frequencies* becomes a greater problem still. Hence there is a good motivation to move traditional analog functions into the digital domain. One approach to addressing this issue is to adopt a software-defined radio (SDR) approach. With SDR, the chipset can be reconfigured in software any number of times to support the modem or modems required for the current mode of operation. SDR approaches have been used for a number of years in base station and network infrastructure. However, SDR has traditionally imposed a power consumption penalty (as more software is being run to replicate previously power-efficient hardware), making it less suitable for devices. Recently, new vector-based processor architecture designs (which perform a set of operations on a 1D array or vector in one operation) and improvements in the efficiency of code produced by compilers, mean that SDR techniques are being applied successfully to solutions in devices. It is noteworthy that, because of the wide range of requirements in the analog frequency domain, it is considerably harder to meet these with a SDR approach than in the digital domain. Ironically it is the digital baseband domain which is able to benefit most from SDR approaches, rather than the real prize of truly being able to design a radio in software. Nonetheless, SDR is bringing benefits in being able to reconfigure a platform across different markets and drives great economies of scale. Following a recurring theme, a platform approach allows the optimal return on R&D investment, as many different final products can be created from the same essential design. An example of a SDR chipset design is Qualcomm's Gobi chipset range (designed for use in portable computing devices), which provides a wireless modem capability that is able to work on numerous modem standards throughout the world.

4.6 Baseband cellular modem design

4.6.1 Cellular modem chipset: purpose

The baseband cellular modem chipset of a mobile device could be characterized as akin to the engine of a car, which allows the car to perform its fundamental function – movement. In a handset, it is the cellular

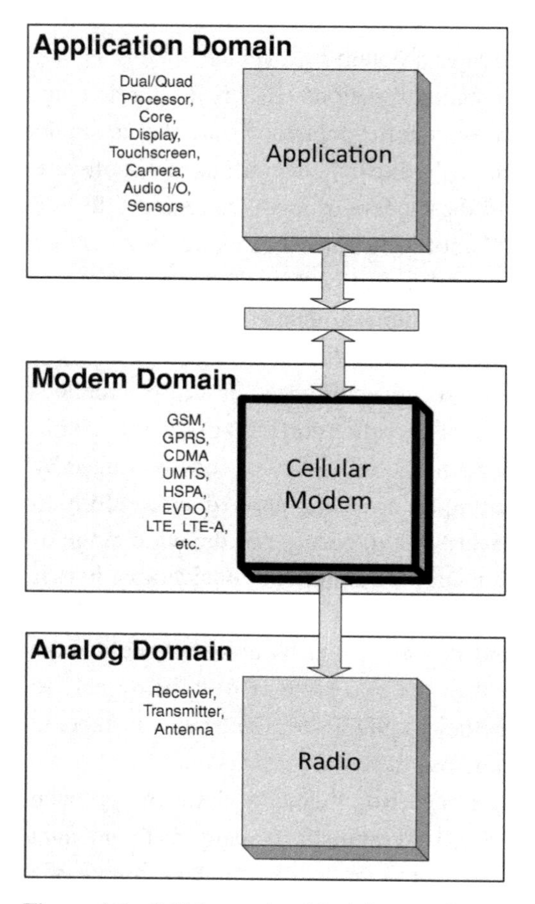

Figure 4.6. Cellular modem block locator diagram.

modem which provides the fundamental function – wireless connectivity. Figure 4.6 indicates the location of the cellular modem within the overall hardware functional block diagram.

It is common to refer to the combination of the cellular modem, radio chipset and associated circuits as "the modem," as a shorthand for everything to do with getting information between the handset and the network. This can be a useful shorthand for designers, whose main focus is on the applications, user experience and overall product design of contemporary devices. A real risk of using this terminology is that of denigrating the *modem* to the status of a peripheral, as if it were a dial-up modem of the 1980s, and, in so doing, under-estimating the centrality of "modem functions" to handset design.

4.6.2 Cellular modem: function

As a generic description, the core platform functions of a cellular modem can be summarized as follows.

- To provide a precise timing framework to ensure that information received from the transceiver block (normally just referred to as "RF," i.e. *radio frequency*) and sent to the RF is acted upon at the correct point in time. This is a critical capability in ensuring that the mobile handset transmits and receives only at the precise times mandated by the cellular standards.
- To provide an information flow framework to ensure that functional blocks within the chipset itself, as well as functional blocks external to the chipset, are able to communicate information to each other. For each wireless interface (e.g. 3G, WiFi, etc.), data and control information need to flow between different functional blocks – for example, incoming voice data received via 3G may need to be re-routed via a Bluetooth interface to an external wireless headset.
- To provide a control framework to ensure that processing modules can signal control instructions to other functional blocks, both within and external to the chipset. For example, the cellular modem communicates with other subsystems such as radio and power management.

4.6.3 Cellular modem: architecture

To the framework are added processing cores, which may be microprocessor-based or DSP (digital signal processor)-based, and also dedicated processing modules (such as a video encoder–decoder or 3D graphics module).

What is the difference between a DSP and a microprocessor?

DSPs are special-purpose processors that have specially designed instruction sets that allow complex signal processing functions which would take multiple steps to achieve with a standard general purpose microprocessor, to be performed as a single operation. The instruction sets and chip architecture of DSPs are specifically optimized to execute mathematical transformation on digital signals.

Microprocessors are general-purpose processors which can perform a set of operations, each of which is comparatively simple, but which can be combined in many different ways to provide considerable flexibility. This flexibility allows any type of software function to be performed, ranging from low-level hardware control through to high-level user interface software. The instruction sets of microprocessors are specifically optimized to execute decision-making logic very efficiently, as exemplified by control software.

The processing modules contain implementations which are specific to the wireless standard being supported. In general, these may be implemented in hardware and software, though the direction of travel is always towards software, as software may be changed or upgraded as required. Hardware processing is preferred where an advantage such as lower power or lower cost can be achieved. Hardware processing requires a very stable and proven implementation, as the cost of changing or upgrading hardware is prohibitive when compared with that of making alterations to the software. With the increase in the number of modem standards being supported by a single chipset (e.g. legacy 2G, 3G and LTE), there is a case for stating that it is becoming more cost-effective to support configurable modem software which can be used with multiple standards, as opposed to continuing to add further hardware processing modules.

4.6.4 Cellular modem: design issues

The time required to design a new cellular modem and get it to market in a commercial handset is typically around three years. By "new," what is meant is a chipset which supports a major step in modem capability – for example, from 2G to 3G, or 3G to LTE.

As the handset market has developed, we have seen handset models changing every year or every season. Clearly then, there is a big disconnect between the design cycle time of a chipset (typically two to three years) and the design cycle time of a handset (typically six to 18 months) that uses such a chipset. Given this time disconnect, it is therefore essential to achieve longevity from the chipset design to allow a range of handsets to be designed, powered by those chipsets. The need to support multiple handset designs, often with a number of conflicting design requirements, over an extended period of time, leads to a platform design philosophy for chipsets. A chipset platform design contains a number of core components (such as a cellular modem capability) with the flexibility to create a family of chipset product variants, each of which may make a different design trade-off on cost, time to market, performance and functionality. This platform approach reduces risk and increases the addressable market for the chipset family. From a common chipset platform design, it is possible to derive a number of different chipset variants, and from these it is possible to create a larger number of handset design variants, potentially across a number of different handset manufacturers, into multiple markets globally.

The industry has typically followed a now classic four-stage approach to developing chipsets and handsets for a new generation of wireless capability, which balances a range of design parameters in different ways as the market develops. This approach is summarized in Table 4.3 and now described in further detail.

Stage 1 of this approach is to create a new cellular modem design which is focused on providing modem-only capability. The resultant products are typically USB sticks with a wireless modem capability for use with laptops and netbooks. During the 1990s and 2000s, wireless PCMCIA or PC cards were the resultant product. The market for wireless USB dongles is growing (tens of millions a year globally), although it is dwarfed by the potential for handset volumes which will follow once modem technology is market proven.

With a proven modem-only chipset, Stage 2 requires a chipset designer to provide variants of the chipset family that are compatible with a range

Table 4.3. *Route to market for next-generation wireless standard*

Market stage	Key market characteristics	Likely dominant devices
Stage 1	Time to market Enhanced modem capability Ability to iterate quickly Low volume Low price sensitivity	Modem-only, e.g. USB modem Smartphone with modem module
Stage 2	High performance New functionality Low but rising volume Medium price sensitivity	High-end handsets
Stage 3	Cost reduction Market niches develop Rapid sustained volume growth High price sensitivity	Mid-end handsets
Stage 4	Cost reduction Reduce specification scope Reduced number of chips through greater integration High volume Intense price sensitivity	Low-end handsets

of application processors and peripheral processors as appropriate. This stage allows high-end handsets to be created, which typically also provide new-generation application capability such as advanced multimedia or new user experience paradigms (3D user interfaces, touchscreens and advanced application/OS capabilities are examples from the recent past).

As the volume of handsets rises further with the adoption of the new modem standard, and consumers begin to enjoy the benefits of higher speed wireless connectivity, the market opportunity is created to address the mass market audience with these newer advanced capabilities – this is Stage 3. In order to address this market, other design parameters come into play. In particular, the need arises to reduce the overall cost of a handset design in order to address the mid-range device classes with the improved modem capability. This cost reduction can be traded to a degree with some degradation in functionality or performance, although the design challenge is to ensure compelling functionality at that lower price point. This has been the story with 3G technology, where chipset manufacturers have focused design integration on cost reduction in order to achieve reach into the high-volume mid-range devices. Eventually, as the modem standard becomes very mature, and the market requirements for low-end handsets emerge, it becomes possible to drive down cost further by removing flexibility from the chipset and by integrating functions further onto the chipset – this is Stage 4. In this market, the chipset functionality is much more fixed – flexibility has been traded for lower cost.

4.7 Mobile application processor design

4.7.1 Application processor: purpose

Application processors provide significant horse-power for accelerating multimedia capabilities such as audio, video, graphics and speech, as well as controlling most of the peripherals of a high-end device – touchscreen, stereo audio, sensors (e.g. motion sensor), memory systems (on-board memory and external memory card), etc. Figure 4.7 indicates the location of the application processor within the overall hardware functional block diagram. Chip manufacturers may also choose to add discrete hardware to an application processor solution in order to offload highly computationally intensive operations such as rendering of 3D graphics or high-definition video. For the same performance, discrete hardware provides the key advantage of lower power consumption compared with a processor executing software from memory. However, hardware presents potential disadvantages of consuming silicon area, and of being inflexible – dangerous if standards are not finalized.

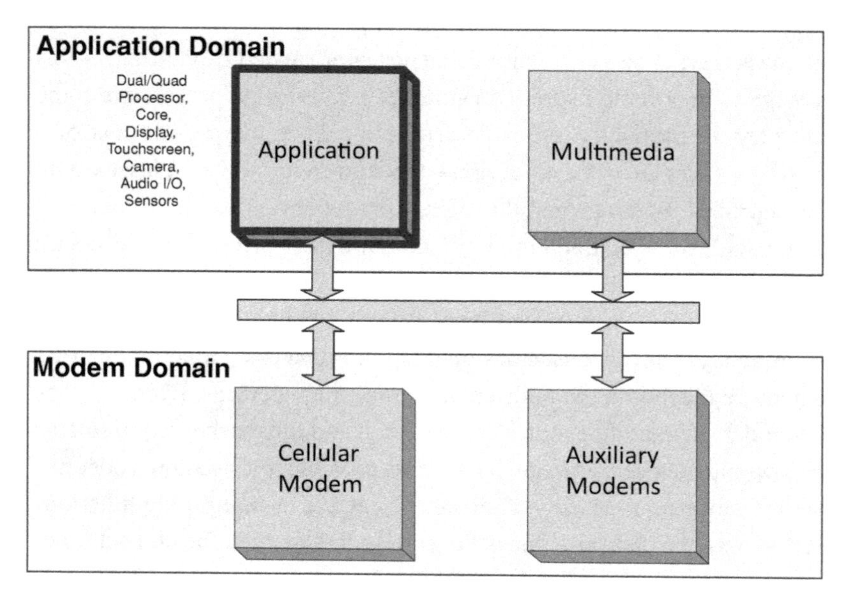

Figure 4.7. Application processor block locator diagram.

4.7.2 Application processor: function

A key function of an application processor is to provide high-performance capabilities for the capture, storage, transmission and playback of multimedia content. Raw data – for example, video frame data or music samples – requires significant memory storage, and hence many algorithms have been developed which compress the data to a more manageable size for both transmission and storage. A *coder* is a function which transforms raw data into a more efficiently coded form. A *decoder* is a function which transforms coded data back into raw data. Combined, these functions are known as a *codec* (coder–decoder). An example is an MP3 codec, which can code from 1.4 Mbit/s for CD stereo music to 192 kbit/s in MP3 format. In addition to MP3, other examples of codecs include AAC for music, MPEG4 and H.264 for video, 1080p for HD video and AMR for speech coding. These codecs may be implemented in software or hardware, but require significant processing power (measured in MIPS – millions of instructions per second), both to read and write data from/to memory, and to perform coding/decoding operations.

The application processor is also responsible for interfacing to an increasingly large number of peripheral devices such as display, touch-screen controller, camera, audio, external memory, USB, sensors, etc. Each of these devices requires additional processing capabilities. These may be provided by the peripheral component supplier as a separate controller chip, or may be integrated onto the application processor chipset. At a minimum, the application processor provides data busses and control lines in order to interface to these peripheral components. As each new peripheral becomes mainstream in the market, the peripheral controller part of a peripheral is highly likely to be integrated onto the application processor, providing reductions in chip count and overall system cost for an OEM.

4.7.3 Application processor: design issues

Contemporary processor cores for application processors focus on optimization at the instruction set level. This means that programming instructions for multimedia operations can be accelerated by translating these higher-level instructions directly into purpose-designed instructions supported by the microprocessor core.

For a number of years in the early 2000s, designers of baseband processors sought to accommodate on-board multimedia functions, with the objective of avoiding the need to add cost to the handset BoM resulting from the addition of a secondary processing chipset. An equally important consideration was that many manufacturers of baseband processors did not have an application processor capability in their armory, and hence there was concern around some of the value of the baseband ebbing away towards a competitor's application processor. However, with the rise in high-performance video and audio codecs to support music and video streaming, download and playback, along with increasing demands of next-generation wireless standards, there was significant technical merit in separating out these two rather different functions of a contemporary device design. In addition, the rate of change in the application domain is significantly more rapid than the time it takes to develop and deploy a new modem standard.

With volumes of smartphones rising (around 500 million units in 2011), and levels of integration increasing (thus reducing the average selling price), there is a strong trend at the time of writing to move multimedia functions out of the baseband processor to consolidate "non-modem" features in the application processor. This approach includes moving the voice codec functions (once a preserve of the baseband DSP) out of the baseband and into the application processor. This design partition approach allows for a simpler design for the architecture of the baseband processor and a relatively standard set of interface points between the baseband processor and the application processor.

Another significant issue is the design cycle time for an application processor chipset, which is much shorter than that for a baseband processor. This is because there are not the same complexities of needing to develop and validate communication systems which are standardized, yet need extensive testing in the field as well as the lab, ahead of commercial product. Application processors are currently seeing large market growth due to the rapid uptake of touchscreen-enabled smartphone devices, and this growing market share and rate of change also encourages a "features arms race" between the chip manufacturers, in itself leading to shorter cycle times.

4.8 Multimedia processor design

Until the early 2000s, mobile handsets were primarily talk and text devices, and the only "media processing" requirements were for the speech codecs which digitized speech frames and encoded them in a highly efficient format for encoding on the air interface, and vice versa. With the advent of early mobile data services such as WAP and MMS, and the addition of digital cameras to mobile handsets, the requirements for multimedia processing increased dramatically. Within a space of just a few years, handset designs evolved from being able to display static black and white bit map graphics, to being able to stream or download a range of audio and video multimedia, including MP3, AAC, MPEG4 and H.263. Camera companion chips offloaded processing requirements from the main handset processor to support image capture and manipulation and various special effects. With the advent of execution engines such as Java for mobile devices, requirements also increased rapidly for 2D and 3D graphics processing capabilities to support a new generation of mobile game applications.

As requirements became established, it also became clear that high-performance multimedia – achieved with a low power budget – was critical for consumer uptake, and thus the market conditions were right for a new breed of separate mobile media processors, now termed graphics processor units (GPUs). GPUs are systems on a chip solutions – with their own CPU, DSPs, memory, power management, discrete hardware and display engines.

GPUs are subject to all of the same design constraints as other chipsets for mobile – size, cost, performance, power consumption, etc. In addition, GPUs have complex power and thermal management issues to contend with. The nature of multimedia processing is that there are very high levels of peak performance compared with average performance requirements. The higher the processing performance required, the more current is required, but also the greater the effects of lost energy through heating effects. Modern designs offer the ability to scale peak performance, to achieve the appropriate design trade-off between performance and battery life, and designers place particular emphasis on reducing current leakage, which is a major source of thermal heating through lost energy.

GPUs may be provided to the market as a stand-alone chipset, or may be provided as a further level of system on a chip along with the application processor capability.

4.8.1 Power management and analog design

The role of a power management chip is to optimize the use of power in the overall handset design. It achieves this by providing programmable power supply lines, with the ability to control the ramp-up and ramp-down sequences in a mobile handset. Various power modes are controlled by the power management component, such as active, standby, idle, sleep and charging. The power management chip is also responsible for providing battery charging functions to the battery, which involves adjusting the current supplied to the battery over time to reflect the chemistry and charging characteristics of the battery.

When the handset is in sleep mode, a 32 kHz crystal oscillator is used to time how long the mobile handset should sleep before waking up to listen for paging messages from the network (for incoming calls, requests for location updates, etc.). The power management component interfaces to this sleep circuitry, and wakes up the required handset modules as and when required.

There are a range of other analog design-related components on the handset. These include converters between the digital and analog worlds – digital to analog converters (DAC) and analog-to-digital converters (ADC). These converters are used to interface to analog components such as keypads, backlights, microphones, speakers and so on. Depending on the specific chipset design, a variety of these analog functions may be integrated on a chipset, possibly along with the power management component.

4.9 Peripheral component design

A modern smartphone has a large number of peripherals. Examples include touchscreen, camera, digital audio and video, external memory, USB and a range of sensors.

The majority of these additional components will require additional processing capabilities to perform their functions; for example, the USB port needs a USB controller to capture and send data, the touchscreen needs a touchscreen controller to convert changes in capacitive charge into a recognition of different finger placements and gestures (such as pinch or zoom). There is a constant vying for position and opportunities for differentiation in play between the suppliers of peripherals – who wish to make their components more sophisticated and hence more valuable – and the suppliers of application processors – who wish to maintain and build the value of their key component. This vying for position allows for a number of different possible solutions – for instance, one handset design may have a separate USB controller and audio processor, whereas another may have these functions integrated onto the same application processor core, or perhaps integrated into a single peripheral chip. The general market trend is for peripheral component manufacturers to lead the initial waves of innovation with a high-value bespoke solution which can easily be added on to existing designs. As initial design wins lead to validation of the market opportunity, and the market begins to become established, volumes rise, and the innovators are able to invest further to improve their designs. With a more established market, the application processor manufacturers begin, in parallel, to add better integration for the peripherals and seek to acquire some of the value from the peripheral component supplier by adding discrete peripheral controllers to their designs. Meanwhile, the more nimble peripheral manufacturer is either performing their own levels of further integration of multiple peripheral controllers onto one chip, or alternatively they have moved rapidly on to innovate in a new area.

4.10 Conclusion

Hardware component design for mobile handsets is driven by the key design parameters of cost, performance and low power. Advances in semiconductor technology continue to permit increasing levels of integration, broadly following Moore's Law. This technology trend results in increases in functionality and performance for reducing cost and size. Because improvements in battery technology are modest at best, innovative techniques at all levels of design which permit use of the limited available energy as efficiently as possible remain paramount. As complexity levels continue to increase, a holistic system view of the whole handset design is ever more essential to achieve the optimal trade-off of design constraints. With a complex ecosystem, including hardware and software component suppliers, OEMs, operators, content and application providers, those organizations able to take a complete "antenna to application" view of handset design are much more likely to be successful.

In the next chapter we examine the structure and complexity of handset software, and consider a range of different design issues and approaches, depending on where in the "software stack" from low-level device drivers to the latest trending app we apply our focus.

5 Software design

In this chapter, we examine the major software components of a mobile handset and the key design issues which affect their design. The Motorola DynaTAC, the world's first portable cellular handset, first announced in 1973, required minimal software, and utilized a simple microcomputer built with fewer than 2000 transistors. A modern smartphone, such as the iPhone 4GS, supports 16 Gbyte of flash and 256 Mbyte of RAM, and could have a billion transistors to perform all of the functions of the device. Given the range of software complexity over time, and the range of device types from a basic ultra-low-cost handset to a high-end tablet, we shall focus on core software elements which exist in various forms in all modern handsets, from ultra-low-cost handsets through to smartphones, and we shall define a model of software structure and its evolution which can be matched to these different device types. This approach permits us then to draw out the key design issues and how these have changed – or not – over time.

For explanatory purposes, we shall divide handset software into three broad groups, each of which has its own distinctive design issues:

- application software design – software related to the provision of applications and services to the user of the device, including underlying OS capabilities where applicable;
- protocol stack software design – software related to the exchange of messaging between the handset and the network in order to provide a set of mobile communication capabilities;
- physical layer software design – software related to the encoding and decoding of data via the air interface, and the control of associated hardware.

After examining each of these software areas and associated design issues, we shall conclude with a review of a set of broader design issues related to mobile operating systems, used in smartphones, and mobile execution environments, used in feature phones.

5.1 Application software design

5.1.1 Application software – purpose and function

The application software is responsible for all aspects of interaction with the user. Until the mid 1990s, application software was centered around core functions of telephony – keyboard input for making and receiving calls, acoustic and alerting controls, dialing aids such as last number dial, phonebook, etc., and display output consisting of perhaps a couple of lines of text and a few icons. Although conceptually simple at first, handset application software has always been challenging due to the typically severe constraints on available memory (both code and data space) driven by memory costs and an acceptable BoM for the current stage of market maturity. Until relatively recently, application software has coexisted with the modem software which has strict real-time requirements, and thus the application software has needed to be structured to ensure that it cannot monopolize processor resources and "lock out" lower-level software from performing these critical real-time tasks.

Nonetheless, looking back, the handset application software of (say) the twentieth century had a modest set of requirements compared to the contemporary mobile computers which increasingly perform most of the functions of a desktop computer, yet – as always – with the classic handset design constraints of power consumption, BoM and form-factor to manage against ever increasing demands on performance and functionality.

As well as managing all aspects of the visible user interface – what appears on the screen, how the keys or touchscreen allow navigation through menus, selecting options and launching handset features – the application software also has to interface to the other software modules, such as the protocol stack (modem) software and peripheral and device software to manage everything from the backlight to the GPS receiver. As discussed in Section 4.5.3 on chipsets, for lower-cost

Table 5.1. *Application software evolution*

Stage	Common terminology	Mental model of what the software does	Distinguishing software architecture characteristics	Iconic phone referenced in Chapter 1
Stage 1	MMI "man–machine interface"	Means to manage a limited number of events between the user and the phone – e.g. call events, buttons, knobs, alerts, informational messages	Each function can be viewed as processing an event (incoming call, button push), which may optionally result in outputting information to a screen, sounding an alert, etc.	Motorola DynaTAC 8000X
Stage 2	Menu-driven user interface	Means to select from an increasing range of possible functions – the menu-driven user interface	Each function can be viewed as a "screen" of the user interface Software architecture needs to manage a large but definable set of screen transitions, with just one particular screen active at any time	Nokia 2110

Stage 3	Execution environment	Means to develop, install and execute third-party applications such as games, normally with no or limited ability to switch between applications, as the execution environment itself runs as one or more tasks under a real-time kernel	The software consists of a number of discrete applications which the user chooses; each application can make use of a common set of software functions such as wireless access, multimedia processing, location awareness, etc.	Sharp GX10
Stage 4	Operating system	Means to develop, install and execute third-party applications, each of which has its own operating system resources such as memory, processing time, etc., allocated to it There is the ability to switch between different running applications, and applications may continue to run "in the background" for periods without interaction with the user required	The software is organized as a computer operating system with supporting run-time environment, software libraries, and the framework which provides the APIs which applications communicate with	iPhone

(and older-generation) designs, all of these functions coexist on the same processor core. With more capable feature phones, and all modern smartphones, the application software and the modem software are segmented between different specialized processors.

5.1.2 Application software – design

Today's smartphones are sophisticated connected devices with a rich application environment. As we have seen in Chapter 1, until approximately 2000, mobile phones were primarily voice-centric devices, with the user interface focused around telephony functions specified by standards bodies. On this journey between a phone without wires and the multi-purpose connected device of the present day, there has been the need for several radical re-thinks of what the purpose of application software is, and therefore how to construct the software architecture to meet emerging needs and future trends. Following a parallel, yet distinct, path has been the evolution in user experience input and output mechanisms, ranging from physical dials through to contemporary touchscreens.

We believe that, up until the time of writing in 2012, there have been four stages so far in the evolution of application software, which are summarized in Table 5.1.

Despite these very different evolutions of application software, we believe it is possible to identify five discrete blocks which are present in each of these evolutions, albeit simpler in the earlier stages and more complex in the later. These blocks can also be identified in the full range of devices from ultra-low-cost handsets through to the most sophisticated smartphone. The five blocks are illustrated in Figure 5.1 and Table 5.2.

The following sections provide a description of each of these blocks, along with a specific case study example around the Android operating system, set as boxed text.

5.1.3 Applications

An application may loosely be defined as a software program which is perceived by the user as performing an overall high-level task or function.

Table 5.2. *Application software architecture*

Block	Purpose
Applications	Software programs which provide a particular utility to a user – e.g. game, browser, navigation service, phonebook, dialer
Application framework	Software which allows multiple applications to access a common set of shared functions or libraries, termed middleware
Middleware	Set of software functions which may need to be used by multiple applications such as making a voice call, sending a short message, accessing an Internet connection, playing video content, providing 2D and 3D graphics
System kernel	Software which manages access to shared resources such as processing time, memory and device drivers, and the sequencing of software execution such as hardware interrupts and the scheduling of software processes or tasks
Device drivers	Provides control of hardware peripherals and exchange of data between the peripheral and handset software

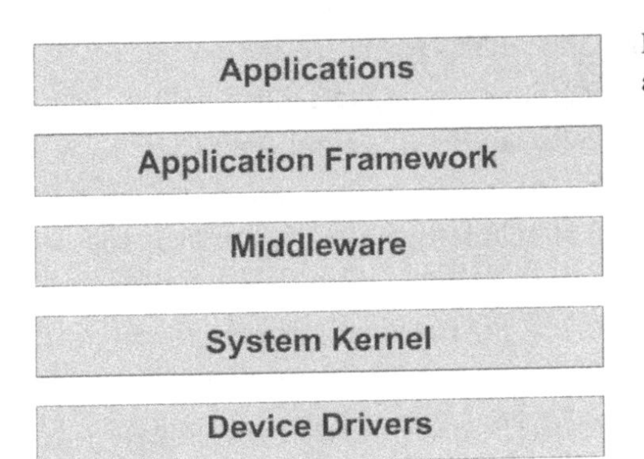

Figure 5.1 Application software architecture.

Examples would include a web browser, a media player, a camera application, a file system or an alarm clock. However, what about standard phone features such as the phonebook, making and receiving phone calls, sending and receiving short messaging and so on? Until the advent of mobile operating systems, these functions had been developed as integral software of the handset by the OEM, and this is what still exists in most mass-market phones today – at this level the phone needs to act more like an appliance than a computer. However, this is largely a matter of user interface presentation, and "under the hood" most smartphones now provide a range of messaging functions via a discrete messaging application, or phonebook functions via a discrete phonebook application.

> Android includes a set of core applications as standard, including browser, e-mail, messaging, calendar, maps, contact management, etc. All applications in Android are written using the Java programming language.

Whatever the function of the application, it needs to communicate with the kernel and its middleware through a set of APIs. For third parties to use these APIs, they need to be well-defined, not change too often and preserve backwards compatibility wherever possible (old functions continue to work even when new functions are introduced). In addition, developers need software tools called SDKs (software development kits) which allow them to create their applications and test them, initially on a PC platform, and later on real handsets. When testing on a PC platform, the SDK must provide all of the same API calls as are found on the handset software, and provide a simulation environment which allows the code to run as if it were interacting with the actual handset software. Debugging of an application on a PC environment is easily an order of magnitude more rapid than seeking to diagnose an error on a handset, where often very specialist debug and diagnostic tools are required.

Applications may be developed by a third party, and then, working with the handset designer, be integrated with the handset software and shipped as part of the software executable of the handset. On feature phones and low-end phones, with simple application platforms, this process could take many weeks or months of integration and testing, as the third-party application is part of the final handset executable software. Increasingly with more advanced application platforms and mobile operating systems, there is a much cleaner and clearer distinction between the application software code and the platform services and APIs which provide the underlying capabilities.

As well as developing and testing their own applications, there is a need to ensure that third-party developers execute their code in a manner which does not interfere with the overall performance of the handset software – and in the worst case cause a software system failure or loss of data. To mitigate these risks, there are a number of test and certification regimes for handset applications. Many of these are run by the handset platform provider or an independent test house authorized by the platform provider. This application certification process ensures the quality of the software executing the application, although the process is not generally to determine the intrinsic merits of the marketability of an application. Marketability and a sales price for the application are agreed with the provider of the download service, which would normally be the network operator, though increasingly it is the platform provider – such as Apple or Google.

Increasingly, applications are written and then downloaded to the handset over the air, or through "side loading" via a serial interface such as USB.

Developers of third-party applications require considerable support from the provider of the software platform. Although third-party ecosystems are beyond the scope of this book, the diagram in Figure 5.2 illustrates a number of the main types of support services that a developer needs and expects from the provider of a software platform such as Android.

5.1.4 Application framework

In this section, we cover the "Application Framework" block of application software architecture, which is the second block indicated in

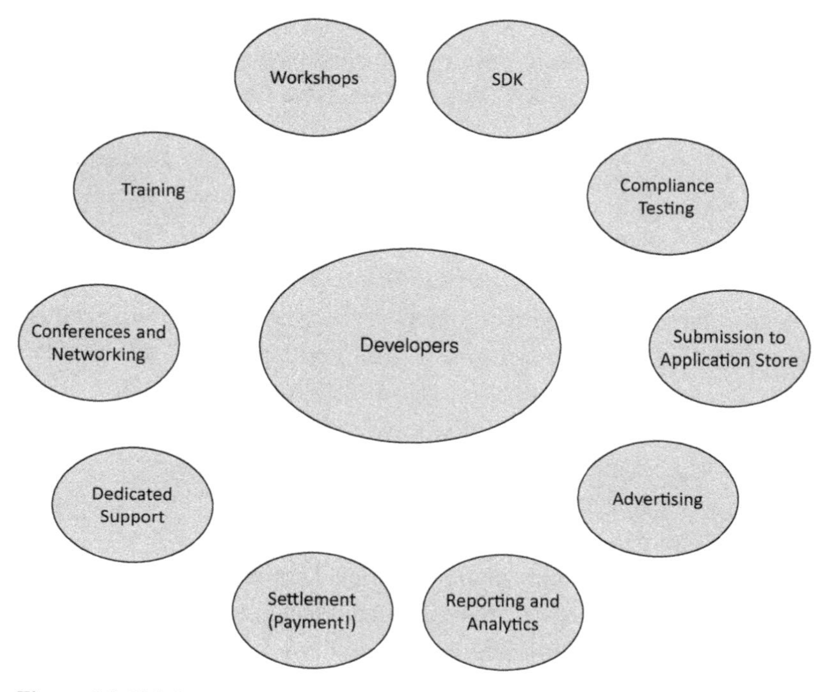

Figure 5.2 Third-party ecosystem.

Figure 5.1. An application framework allows the re-use of common software (i.e. middleware) between applications.

Beyond the simplest user interface requirements, some form of software framework or structure is required. For instance, how do you manage being three levels deep within the menu structure when a text message arrives or the battery low event occurs? How do you manage overlaying an informational or error message on the screen, and ensuring that when the message is removed the screen that was shown previously is re-displayed?

These issues are addressed by use of an application framework. For traditional handset architectures not using an OS or execution environment, these are totally proprietary and nearly always created in-house by the handset manufacturer's software team. There are no rules or standards as to how to create such a framework, but an approach needs to be developed based on sound software architecture and design principles. In the past, a number of handset manufacturers have suffered significant product delays and escalating software development costs, as their application framework design was unable to scale to a new wave of feature requirements, and a major re-design of the software architecture was then required.

In Android, the application framework provides third-party developers with the ability to develop their own applications, using the same set of APIs used by the core applications shipped with Android. Through the framework, developers are able to access many of the underlying capabilities of the phone. For example, this includes accessing the device hardware, accessing location information, running background services, accessing aspects of the core user experience such as notifications and the status bar and so on. According to the Android developer website: "The application architecture is designed to simplify the reuse of components; any application can publish its capabilities and any other application may then make use of those capabilities (subject to security constraints enforced by the framework). This same mechanism allows components to be replaced by the user."

The framework provides a consistent approach to creating the user experience, including lists, grids, text boxes, buttons and so on, for accessing data from other applications (such as contacts from a contacts manager) and for managing the execution of an application such as starting, stopping, pausing and ending.

5.1.5 Middleware or application libraries

In this section, we cover the "Middleware" block of application software architecture, which is the third block indicated in Figure 5.1. Middleware, or *application libraries*, are somewhat elusive terms, that refer to software which is neither low-level (close to the hardware) nor high-level (close to the user) – but somewhere in the middle. The reason we have middleware at all is to ensure that functions and operations that need to be performed by more than one application are only designed once and are present once in the handset software. Middleware is a disparate set of not very glamorous, yet critically important, software functions which enable applications such as browsers, scheduling calendars and media players to achieve their purpose. Significant development effort is required to construct and maintain middleware in a suitable modular yet efficient form, in order to stand the ravages of time as hardware platforms change and evolve on the one hand, and applications and user interface models (e.g. the emergence of touchscreens) stretch and expand software requirements on the other.

> Android provides a set of Middleware libraries, written in C/C++, which are then exposed to application developers in Java through the Android Application Framework. Example capabilities include: recording and playback of audio and video; image rendering; access to the display; managing 2D and 3D graphics; font rendering; relational database support (SQLite).

Earlier, we defined four stages in the evolution to date of application software, which followed in progression as the capabilities and requirements of handsets have increased. In the late 1990s, the industry started to move between Stage 2 (menu-driven user interfaces) and Stage 3 (execution environments) following the evolution of handsets from telephony devices towards multimedia and Internet-enabled products. Within this transition, the middleware requirements increased very dramatically very quickly, and required detailed knowledge in new areas such as multimedia codecs, web browsers and Internet protocol stacks, and 2D and 3D graphics. Middleware became “bloatware,” and the economics and practicalities of handset manufacturers creating all of this software in-house no longer made sense. Many third-party software companies emerged to license software components to handset manufacturers on a more economic basis. Just a few example companies at this time included

Access, Obigo, Openwave and Opera for web browsers; Esmertec and Sun Microsystems for Java virtual machines; Magic4 for messaging clients; Packetvideo for codecs; Certicom and RSA for cryptographic software; In-Fusio for games engines; Tegic and Zi for predictive text entry; ART, Nuance and Voicesignal for speech recognition; Action Engine, ActiveSky, SurfKitchen, and Trigenix for on-device shopping portals and so on. As might quickly be observed, this became a logistical nightmare for handset manufacturers and platform providers to manage, and the economics did not fundamentally work, as there were many software companies, each with a business plan predicated on obtaining a decent per-unit royalty on each handset which shipped containing their software. However, from the handset manufacturer perspective, middleware was costly, was rarely a market differentiator, was hidden from their customer or end user and, because of the complex integration involved, added significant project risk to their handset programs.

Taking the long view, the poor economics of this situation created the business need for a new approach. One such approach was growing interest in the use of open source software, which offered many common software components for low (free) initial licensing costs, as well as the emergence of Linux as a contender for an underlying operating system for which much open source software was readily available. Concerns around open source included quality assurance, lifetime costs of maintaining the software and some types of licensing terms which required any software which used or interacted with the open source software to be made freely available to the wider community.

In the mid "noughties," significant industry players such as Nokia and Motorola began to build certain product ranges based on open source platforms, such as Nokia's Maemo and Motorola's JUIX platform. A new platform dilemma was also emerging between the embedded software world (which included Linux), focused around the C/C++ development languages, traditionally very well suited for writing memory efficient code with good alignment to underlying hardware operations, and the emergent "online world" (including gaming), which was largely built around Java and HTML technologies. From very early on, RIM – maker of the BlackBerry devices – successfully developed their own Java-based application environment, which interfaced to underlying protocol and modem software, written in C/C++.

The stage was therefore set in 2008 for Google to enter the market with a new mobile operating system – Android. Android is built on a Linux platform, utilizes much already available open source software, and then builds on top of this a Java-based environment which provides the application platform within which applications run. By bringing together all of these elements into a single platform, and making the software available on open source terms, Google created a platform which provided a new economic model for handset manufacturers to reduce software R&D costs, whilst being able to bring to market a new generation of feature-rich, highly capable smartphone devices. Through close work with semiconductor companies, Google is increasingly able to position Android into sections of the feature phone segment, providing further economies of scale for all concerned.

5.1.6 System kernel

In this section, we cover the "System Kernel" block of the application software architecture, which is the fourth block indicated in Figure 5.1. The system kernel is software which manages access to shared resources, such as processing time, memory and device drivers, and the sequencing of software execution, such as hardware interrupts and the scheduling of software processes or tasks. Aside from smartphone operating systems, most system kernels are proprietary real-time operating systems (RTOS) provided either by chip manufacturers or specialist providers such as Nucleus and OSE – both of whom have product shipments measured cumulatively in the billions. The most important function of these RTOS is to provide a scheduling method for software tasks and a framework for handling hardware interrupts and executing interrupt service routines.

5.1.6.1 Task scheduling

A task is a software program which requires use of shared system resources such as processor time, memory, timers and device drivers, and has the ability to communicate with other tasks in order to achieve a higher-level objective. Multiple tasks exist together and need to communicate together in order to achieve the functions of the software system – in our case, a mobile handset. Tasks may either be allocated processing time on a time-share basis, or, more typically for embedded systems, continue to execute until they relinquish control back to the task scheduler.

Device drivers which need to read or write data to/from hardware are normally written as interrupt service routines (ISRs), which means that when there is data to transfer, these routines can interrupt the currently running task, perform their function, and then relinquish control back to the task. Normally these interrupts are managed by the microprocessor, which means that the scheduler, which is a software program itself, may have no involvement in managing these interrupts.

Tasks may be given different levels of priority. Typically, tasks which have to provide interactions with hardware in real time are assigned a much higher priority than application-oriented tasks which are less time-critical. This explains the symptom on desktop machines whereby applications may appear to freeze or go on a "go slow" whilst the processor is busy servicing other lower-level tasks.

Android includes a set of its own libraries which provides most of the functionality available in the core libraries of the Java programming language.

Every Android application runs in its own process, with its own instance of the Dalvik virtual machine (VM). Dalvik has been written so that a device can run multiple VMs efficiently. The Dalvik VM executes files in the Dalvik Executable (.dex) format, which is optimized for minimal memory footprint. The VM is register-based, and runs classes compiled by a Java language compiler that have been transformed into the .dex format by the included "dx" tool.

The Dalvik VM relies on the Linux kernel for underlying functionality such as threading and low-level memory management.

> Android relies on Linux version 2.6 for core system services such as security, memory management, process management, network stack and driver model. The kernel also acts as an abstraction layer between the hardware and the rest of the software stack.

5.1.6.2 Interrupt service routines

When a hardware interrupt event occurs, the software currently executing is stopped, and an interrupt service routine is executed to process the interrupt. Typically, this involves reading data from a hardware register into an area of memory, and vice versa. Interrupt service routines are used where it is critical to interact with hardware within a very specific period of time. These small pieces of software are often hand-crafted by software designers both to meet the real-time requirements of the hardware, and to release control back to the RTOS as soon as possible in order to allow other software to execute.

5.1.7 Device drivers

In this section, we cover the "Device Drivers" block of the application software architecture, which is the fifth block indicated in Figure 5.1. Device drivers are software programs which manage the direct interaction with a hardware interface or peripheral such as the display, external memory, Bluetooth or USB. Each device driver is different, as it needs to "talk the language" of the specific hardware. However, there is likely to be a level of commonality at the interface between the device driver and other task-based software. This is in order to provide a standard framework which many developers can understand when interfacing to such device drivers.

Device drivers may execute when an interrupt occurs from the hardware, or may run on a timer signal, in which case they poll the hardware to see if there is something to do. In mobile handset design, it is undesirable to have software which polls. This is because, in order to poll, the processor may need to be brought out of a low-power sleep state (if, for instance, the handset is in a standby mode). If on most occasions there is no processing to perform (the reason for polling is to check whether there is anything to do), then current has been consumed needlessly. Much better is to have device drivers which are triggered by an external hardware event (such as a key being pressed by the user, or a local connectivity interface (e.g. USB) having data or control information to pass on from another external system). In the case of receiving information from the air interface itself, the mobile communication standards are designed in such a way that it is possible for the handset to be in a "sleep mode" a lot of the time, with well-defined points in time where incoming information from a base station may be transmitted. As we continue to find, everything in the design of mobile communication systems must strive to ensure that the minimum power necessary is utilized by the handset in order to perform the required functions. This philosophy extends also to the design of the communication standards themselves.

5.2 Protocol stack software design

5.2.1 Protocol stack software – purpose

Protocol stack software is responsible for managing a complex set of control signaling between the mobile handset and the cellular network, in order to provide a set of telephony services to application-oriented software.

A *protocol* is an established and understood method of communication between two entities, which involves the exchange of a number of different messages according to a set of established rules, in order to achieve some objective. The use of the term in telecommunications is making an analogy to the use of the term in the diplomatic field, where communication between governments has developed over the centuries to allow two parties to communicate – even if they don't like each other or are at war with one another! This is an important point, as a protocol assumes that the two entities who are communicating with each other are peers – that is, equal in status. In telecommunications, a protocol is a well understood set of communications that can occur between two peers. From the perspective of a mobile handset, one of these peers is always the handset, and the other is the mobile network.

The term *protocol stack* is well used in the wider telecommunications industry, and refers to a number of different protocols stacked up on top of each other, much like one might have a stack of books or a stack of dishes. Each protocol on the handset is responsible for communicating with its peer protocol on the network. This communication is achieved by an exchange of messages. But how do the messages actually get transferred? If we retain the analogy with government diplomacy, all should hopefully become clear. Imagine that we have government leaders of two different nations who wish to communicate with each other diplomatically. The leaders achieve this communication by asking their diplomats and other government officials to undertake the communication on their behalf – generally it is not the leaders themselves who talk directly with each other. It might be, for example, that the wishes of the leader are communicated to the foreign minister, who in turn works with a department for foreign affairs, who in turn works with a local diplomat working for the other nation. So here we have the idea of two peers (government leaders) communicating with each other, not directly, but through their intermediaries, potentially at a number of different levels of seniority.

So it is with a protocol stack. The highest layer of a protocol stack is akin to a government leader, with the lower layers corresponding to the different layers of government officials and diplomats who carry out the requests of those "more senior" people. Note also that the higher-layer messages could be very direct, whereas the diplomatic messages tend to be couched in a more nuanced language that allows for some difference of opinion along the way, without either nation being seen to disagree officially with the other, a situation that might cause a "diplomatic incident."

In a telecommunications protocol stack, therefore, the direct order might be "make a telephone call to Alice." The lower levels of the protocol stack would then take this message and have a much more detailed conversation with their peer entities on the network side in order to locate Alice, allocate resources for the telephone call, deal with the handset moving between cells, and so on.

5.2.2 Protocol stack software – function

The protocol stack software is responsible for negotiating with the network over a wide range of different tasks. Examples include:

- registering the handset to a particular network;
- authenticating the handset and the user of the handset to the network;
- reporting where the handset is (so that the network can route incoming calls to the handset);
- managing the handover of communications from one cell site to another as the handset changes its physical location;
- managing the process of setting up and closing down circuit-switched telephone calls;
- managing the process of establishing and maintaining a session for the transfer of data packets;
- transmission and reception of text messages (SMS).

5.2.3 Protocol stack software – design

5.2.3.1 The seven-layer model

Given the principles of a protocol stack defined above, there are numerous possible ways to think about how to structure the number of layers and the function of each layer. An issue arises, though, when we want to provide inter-operation between two different communication systems – for example, a telephone system and a computer networking system for the transmission of voice or data. In order to carry end-to-end information across two or more different (heterogeneous) networks, it is necessary to have a form of translator to convert between the information flows and protocols on one network and those on the other. Due to the need to make it easier to connect different networks together, a standard for telecommunication protocols was established in the 1970s by the International Organization for Standardization (ISO) to provide a common reference point. This standard model is known as the Open Systems Interconnection (OSI) seven-layer model – or "7 layer model" for short. The OSI model established an idealized protocol stack of up to seven layers as shown in Figure 5.3.

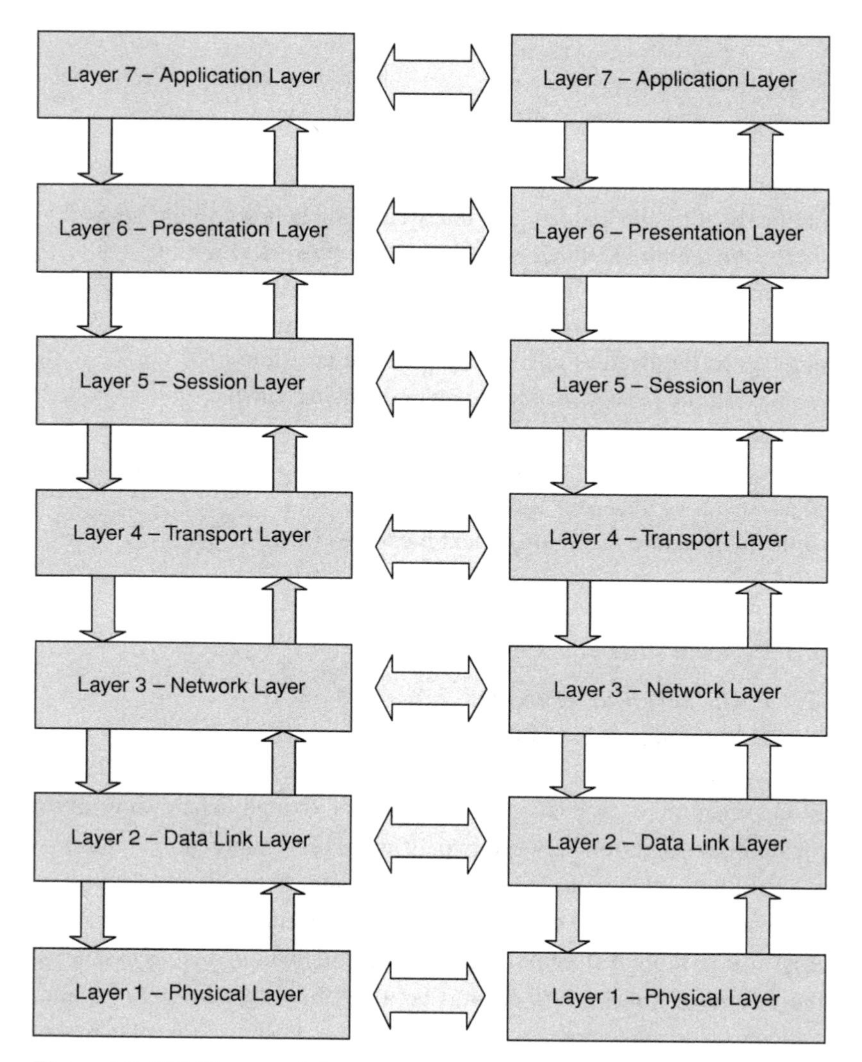

Figure 5.3. ISO 7 layer model.

A key principle of protocol stacks is that a protocol entity can only get a message sent to its peer entity by asking the next lowest protocol stack layer on "its side" to send the message on its behalf. This carries all the way down the layers of the stack, which means that it is only at layer 1 (the physical layer) where actual real transfer of information occurs between

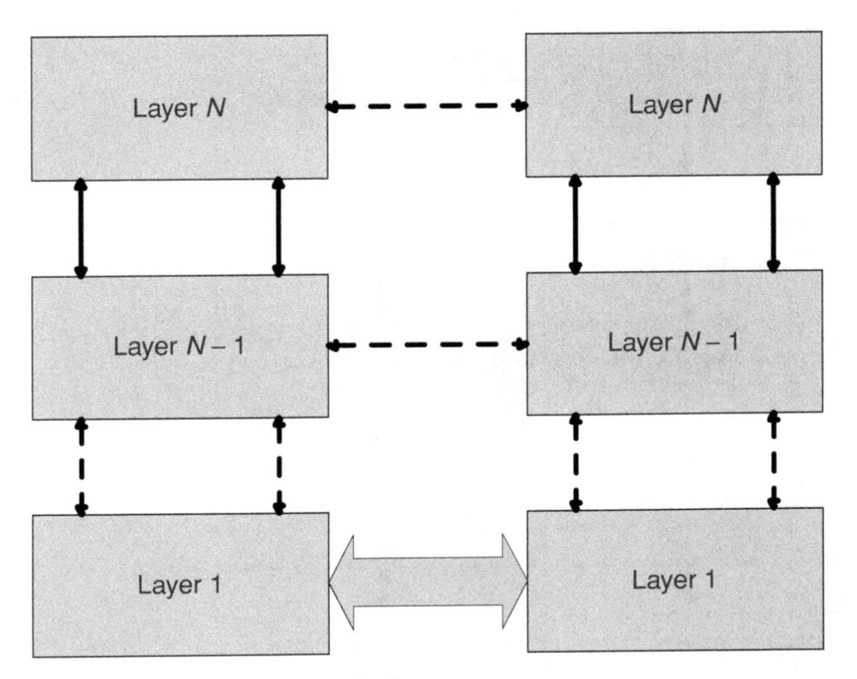

Figure 5.4. Peer-to-peer communication.

the handset and the network. See Figure 5.4. So, for instance, for a layer 3 protocol entity to send a message to its peer in the network, it must send a request to its layer 2 protocol entity to send a layer 3 message on its behalf. The layer 2 protocol entity must then in turn send a request to its layer 1 protocol entity to send a layer 2 message on its behalf to its layer 2 peer. Once the layer 1 entity is reached, then actual communication with the peer entity (the network in our case) can be achieved. In a handset, this is achieved by translating digital bits of information into radio frequency (RF) electromagnetic waves, which are transmitted and received by the network radio receiver. The radio receiver translates back from RF to digital bits, and then goes through various software functions to decode the bits and reconstruct the original message sent by the handset.

Note that a number of layer 1 messages may need to be communicated in order to deliver a complete layer 2 message across the air interface. Some of the layer 1 messages will be "housekeeping" messages which are

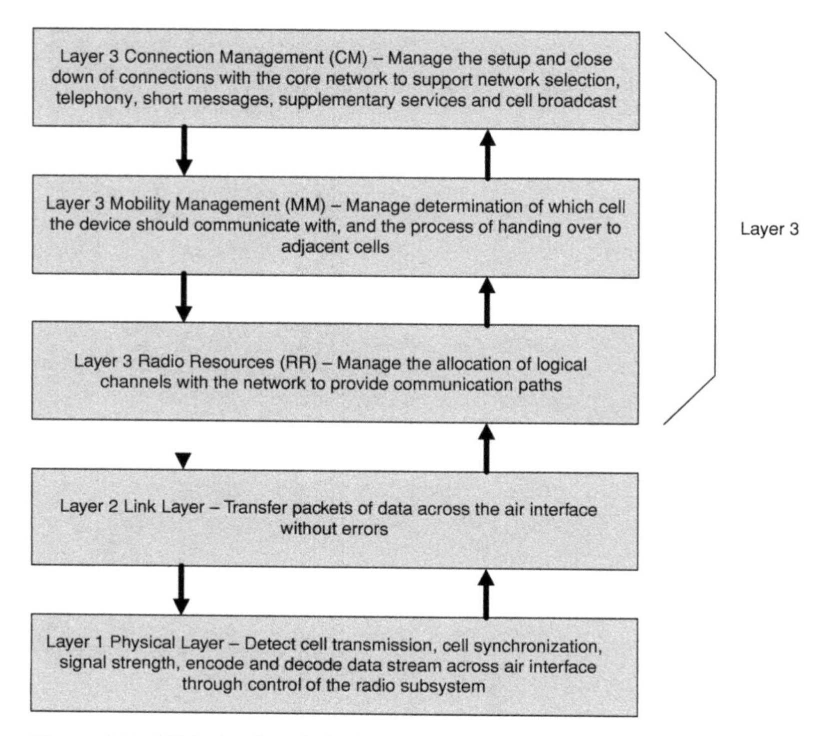

Figure 5.5. GSM circuit-switched protocol stack.

internal to the layer 1 functions and allow the layer 1 entities to perform their functions. Some of the layer 1 messages will indeed contain all or part of the layer 2 message. This principle carries on up through each layer of the protocol stack – so layer 2 messages will be a mixture of internal layer 2 "housekeeping" messages along with messages that carry the contents of the layer 3 message, and so on.

We can now seek to apply this theoretical model to a real-world example, in this instance a GSM (circuit-switched) protocol stack. Figure 5.5 shows the high-level structure of a GSM protocol stack. By contemporary standards of protocol stacks which support both circuit- and packet-switched communications across multiple network standards, this can be viewed as a more straightforward example.

The ISO 7 layer model was first developed with wired solutions in mind, although it has also been utilized very extensively and successfully in wireless communication systems. In the example for the GSM protocol stack, the layer 3 network layer has considerable complexity in order to deal with aspects of mobility and radio resource allocation. As a result of this, the canonical layer 3 as defined in the ISO model is further defined to have three sub-layers as follows.

- **Connection management (CM)** – to deal with the connection issues of setting up, maintaining and closing down a connection. This sub-layer is most like a traditional layer 3 found, for instance, in ISDN, as used for digital wired telephony.
- **Mobility management (MM)** – to deal with all of the issues of the handset moving from cell to cell. In order to achieve mobility, the handset must listen for broadcast signals from both its current cell (known as the serving cell) as well as all the other cells in the neighborhood that it can also detect (known as the adjacent cells).
- **Radio resource management (RR)** – to deal with the requesting and allocation of logical channels from the network to allow for different types of control and data signaling to occur over the allocated frequencies within the radio spectrum.

5.2.3.2 Finite-state machines

From a software design perspective, protocol stacks lend themselves to being implemented using a powerful technique known as finite-state machines, which provide a way of sequencing rigorously the communication between two entities – having received a message from a peer entity, a finite-state machine is able to provide a mechanism to determine the appropriate message to send back to the peer entity, along with the messages to send "up" to the next higher layer and "down" to the next lower layer.

What is a finite-state machine?

A finite-state machine is a system which has a defined number of internal states (colloquially "modes of operation"). The internal state

("state") can only be changed as a result of a new external input to the system. The outputs generated by the system are affected by the change in external input and the current state of the system. Similarly, the state of the system after processing of the external event may be altered, determined by the external event and the current state of the system. The state of the system may also be changed as a result of processing the input event.

Think of a finite-state machine as a "black box," connected to a number of on/off input switches and output lights. By flicking various switches on or off, a different pattern of lights on the output can be illuminated. Each time an input switch is flicked, the black box makes a decision on which output lights to illuminate, based on the new input and the system's current state. Of note is the fact that there is no direct connection between an input and an output, rather the output is a function of both the current state and the external input:

output [1, . . . ,N] = function (new input event, state (current)),
state (next) = function (new input event, state (current)).

5.2.3.3 SIM management

Somewhat related to the protocol stack is the management of the subscriber identity module – or, to give it its more common name, the SIM card. The SIM card contains a set of user data including phonebook entries, SMS storage and parameter settings. In addition, the SIM card plays a central role in the GSM system of representing the digital identity of the user to the network, and of being a key player in the role of authenticating the handset to the network, and vice versa. The SIM card contains a set of secret cryptographic keys, known only to the network operator, and used in authentication ("Is this the expected phone talking to the expected network?") and encryption (encoding speech and data traffic between the handset and the network to prevent eavesdropping by third parties). SIM cards may also store software programs embedded on them by the network operator, which can customize the capabilities of the handset. This capability is known as the SIM application toolkit. Handset software therefore exists to communicate with the SIM card – to read off data files, to request cryptographic functions to be performed, and to support SIM applications which may communicate with the handset software and influence the behavior of the handset, for instance by showing an extra menu screen containing specific operator services which may be activated via the handset user interface.

5.3 Physical layer software design

In this section we provide an overview of the key design issues for physical layer software.

5.3.1 Physical layer software – purpose

The physical layer software is responsible for achieving the encoding, decoding, transmission and reception of digital binary information (bits) between the handset and cellular base stations, for monitoring the status of other channels and neighboring cells, and for achieving the handover process between cells without loss of connection.

5.3.2 Physical layer software – function

The physical layer software orchestrates and controls a range of hardware components such as the RF transmitter and the RF receiver. It provides functions to encode and decode data bits to ensure that the original sequence of bits can be reconstructed at the other end, given that the RF medium may suffer from interference and data loss. It maintains synchronization with the network and manages timings to ensure that bits are encoded and transmitted at an expected time, and it attempts to receive and decode bits at the correct time. It adjusts the transmitter power to optimize the signal strength and adjusts the receiver gain to optimize its ability to receive successfully. All of these functions must be managed so as to minimize the use of power, in order to maximize the life of the battery between charges.

Physical layer software has a major role to play in listening for other channels and cells (cell search), monitoring the signal strengths of neighboring cells to assist decisions about handover (cell measurement), and, during the handover process itself, performing functions to ensure that no data loss occurs.

5.3.3 Physical layer software – design

Physical layer software is divided between functions which are performed on a microprocessor and those which are performed on a digital signal processor (DSP). Both of these processors are hardware components of the baseband chipset, which forms the heart of the telephony capability of a handset.

Physical layer software design is a highly specialized discipline, which requires particular attention to the severe real-time requirements of managing the encoding and decoding of data on and off air, the control of low-level hardware functions and design for power efficiency.

Mobile communication standards have as one of their design constraints the need to optimize the amount of time the mobile handset can be in low-power modes such as standby, in order to maximize battery life. One of the approaches to achieve this is to have precisely defined points in time where the network could send a paging message to the mobile handset to inform it that a signaling exchange is required – e.g. to respond to an incoming call request. This means that, at other times, the mobile handset is able to be placed into a very low-power "sleep mode" where many of the circuits of the chipset are powered down. From a software design perspective therefore, the software is structured such that it can "wake up" when a timer indicates that it is time to monitor a paging channel, perform a set of necessary operations very quickly, and then be able to "go back to sleep" again.

Physical layer software has to achieve sets of functions within very fixed "deadlines." The software may need to process data which has been received from the air interface, make a decision based on the data, and then form a response to the network before the deadline is reached. For example, in the GSM standard, within every 4.6 ms time period, the software has to be able to receive a data block from one channel, transmit a data block on another channel, and monitor one other channel (for measuring signal strengths of neighboring cells). There are no second chances! The software has to be designed to guarantee that operations can finish within a particular timeframe, otherwise data will be lost – at best requiring retransmission, at worst breaking the connection between the handset and the network. Neither of these is acceptable for a commercial product.

Consequently, the physical layer software always runs as the highest-priority software function within the modem processor – all higher-level protocol stack and application functions must be designed to work on longer timeframes (tens or hundreds of milliseconds). Typically, most of the physical layer software will run as interrupt service routines. The processor will receive an interrupt signal from an external source (other hardware or a timer unit), which will cause the processor to pause whatever processing was already in train, and then invoke a pre-determined software routine, known as the interrupt service routine. Because the interrupt service routine prevents all other software running whilst it is executing, this software must be designed to be extremely efficient – colloquially "to get in there, get the job done and get out again." The design philosophy is therefore to do the minimum necessary work in the interrupt service routine (for example, reading data from hardware and storing it in fast access memory), and to offload less time-critical functions, such as unpacking the contents of a data packet, to higher-level software tasks. During one 4.6 ms period in the GSM system, there could be dozens of such interrupts serviced.

In order for the high-priority interrupt service routines to complete their task as rapidly as possible, an important design approach is to minimize the amount of software logic in such a routine. This requires careful software design. The general approach should be to make as many control logic decisions as possible in less time-critical layer 1 software ahead of time, so that, when an interrupt service routine runs, it has a very clear, focused task to perform, without a lot, or any, decision making to perform "on the fly."

In summary, then, it is a fundamental design consideration for physical layer software to be able to perform certain tasks within "deadlines" of time, and the structure and organization of the software must be constructed around these time budgets.

5.4 Mobile operating systems and execution environments

By the end of 2006, Symbian, the supplier of the then dominant mobile operating system, had shipped its software in over 100 million devices. In January 2007, Apple launched their first mobile handset, the iPhone, an iconic touchscreen smartphone running the iOS mobile operating system. In September 2008, T-Mobile USA launched the first ever phone based on the Google Android mobile operating system. In the first quarter of 2011, Android overtook Symbian as the market leading mobile operating system, with an estimated 36 million units shipped, versus Symbian's estimated 27 million, with the total market for phones based on a mobile operating system exceeding 100 million units in that quarter. Also in 2011, Nokia announced that it was building the future of its operating system strategy around an alliance with Microsoft and its Windows Phone operating system, and Google announced its plan to acquire Motorola Mobility – the mobile devices arm of Motorola. The year 2011, then, appears to have been the transition point where the handset industry passed the baton of control and innovation in mobile operating systems and their associated ecosystems to the computing industry, marking another milestone in the development of this fascinating industry.

Modern smartphones and other high-end connected devices have become sophisticated mobile computing devices, which, from a software perspective, rely upon a mobile operating system (OS) to manage the many different functions and applications which such devices run. There is no doubt that the mobile OS is a core component of a contemporary smartphone device. In this section, we look at the key reasons why modern phones need an operating system and we take a high-level view of the main components of such an operating system.

Mobile OSs provide an advanced software environment which enables third-party developers to create software applications ("apps"), which can be downloaded and run on devices, in much the same way as application software can be downloaded and run on PCs. So-called *mobile execution environments* have been achieving a seemingly similar capability in the feature phone segment since 2001, when the first Java-enabled handsets launched in Japan, followed in the same year by the first BREW-based devices in Korea. Due to their dominance in the high-volume mid-tier section of the market, mobile execution environments have shipped in billions of devices since their first market entry – we therefore need to be clear on the differences between execution environments and operating systems.

5.4.1 What is the difference between an OS and an execution environment?

An execution environment provides the ability for many applications to make use of the same common functions, but with only one application doing so at a time (e.g. either play a video or browse the Internet, but not both), whilst an operating system provides the ability for many applications to share the same functions, with multiple applications active at the same time. Both execution environments and operating systems provide a set of software interfaces known as application programming interfaces (APIs), which allow new applications to be written that make use of these interfaces to access the functions of the operating system.

A number of mobile operating systems are available in the market today; the most prevalent solutions are listed in Table 5.3.

5.4.2 What drives the need for a mobile OS or execution environment?

There are three overriding factors which drive the need for an OS or execution environment. These are:

- complexity,
- concurrency,
- community.

5.4.2.1 Complexity

As the number and sophistication of functions in the application space increase, the level of interaction between applications and underlying services increases exponentially. (In the limit, every application needs to talk to every underlying service, and in some cases applications will also

Table 5.3. *Mobile operating systems*

Operating system	Developed by	Comments
iOS	Apple	Operating system for Apple devices such as iPhone and iPad only, derived from Max OS X
Android	Google	Linux-based and available as open source Widely available from a range of handset vendors
BlackBerry	Research In Motion (RIM)	Java-based operating system for RIM BlackBerry devices
Windows Phone	Microsoft	Microsoft has developed a number of operating systems for consumer devices including Windows CE and Windows Mobile Windows Mobile was launched in 2010. Microsoft's partnership with Nokia provides a strong route to market for this operating system
Symbian	Symbian Foundation	The prevalent operating system used by Nokia; until 2010, was globally the OS with largest market share, now overtaken by the rapid growth of Android
WebOS	HP	A new operating system, based on Linux originally developed by Palm, then acquired by HP At the time of writing, its possible future is the subject of hot debate

need to communicate with each other.) The cost of developing, testing and maintaining the software follows a trend related to the number of interactions, rather than the number of new features and capabilities added. Economically then, it becomes ever more expensive to add new features to the product; technically monolithic software architectures which had met the need well in the past face meltdown.

5.4.2.2 Concurrency

Concurrency is the (apparent) ability of a computing system to do more than one thing at once. In practice, with single-core processor systems, the computer is only performing one instruction at a time. However, the operating system provides a software environment which allows the processor resources to be shared between multiple tasks very frequently, providing the appearance of concurrency. Various scheduling mechanisms are established within the field of computer science to manage concurrency, though they are beyond the scope of this book.

5.4.2.3 Community

Opening up development to third parties becomes both an opportunity and a necessity as software complexity increases. It is an opportunity, as it enables many more creative people to have the means to create useful and innovative applications for handsets. New business models become possible as third parties are engaged in the process of creating and delivering aspects of the overall user experience. Consumers have more choice of applications available to them, and only the applications they wish to use need to be loaded onto the handset, rather than the handset manufacturer needing to make a "lowest common denominator" judgment call on which applications to include. It is also a necessity, as handset manufacturers are otherwise unable to keep pace with the demand for new features and applications in an area which typically is not part of their core business of designing and shipping high-volume products. Richness in applications is therefore not ground on which they wish to compete and differentiate, so enabling an "outsourcing model" makes sound business sense.

5.4.2.4 Execution environments

A compromise position is found in the feature phone segment, with the use of execution environments. Indeed, it could be said that feature phones are all about compromise – seeking to achieve just a high enough specification point with advanced features, but at the lowest achievable price point. A feature phone will never be "best" at any particular contemporary feature, rather the objective is to be "good enough" for the majority of users and consumers. To that end, execution environments also seek to deal with the three key issues of complexity, concurrency and community in a more limited fashion.

Regarding complexity, execution environments achieve the objective of "offloading" the development of user applications to third parties to a reasonable degree. However, traditional handset applications, such as phonebook, dialer and browser, are most likely still to be present as native applications (part of the handset OEM's software build). This results in the handset manufacturer carrying a large software burden in managing the complexity of design, test and release of software, now including a range of software from third parties, with their own architectures and APIs. The handset manufacturer has to balance the high cost of software development with the benefit of achieving a lower-cost (BoM) product which can ship into high-volume markets due to its attractive price point.

Regarding concurrency, execution environments typically support just one application being active at a time. Indeed, from a software architecture standpoint, the execution environment itself is normally a task itself, running on top of a low-level OS which is providing scheduling capability for the many different application, communications and peripheral activities of the handset software. In practice, the execution environment is really an application itself, masquerading as something akin to an OS to third-party applications!

Regarding community, execution environments provide a set of APIs for developers to create applications in the same way as an OS. However, due to the constraints on memory and performance, and in some cases the level of peripheral support (e.g. Bluetooth or GPS, or not), not all APIs may be supported on all handsets, which has often led to a high degree of fragmentation. The consequence of this is that developers have effectively many different platforms to develop for, and so the developer market is inhibited from achieving scale, and is encouraged economically to develop less sophisticated applications which can run on the largest installed base of handsets.

5.5 Conclusion

The first cellular handsets required minimal software, with most of the complexity being in electronics. The advent of 2G digital communications was driven by the availability of DSP and microprocessor technologies, which permitted the fast real-time processing of digital information. As processing capability increased dramatically according to Moore's Law, it became possible to add much more sophistication to the application software. Initially this software was seen as an extension of the modem software. As application complexity increased, it became necessary to create new software frameworks, which in the last few years has led to the emergence of a new set of mobile operating systems. Initially these were developed within the traditional handset ecosystem, though now we see the strong emergence of platforms created and supported by heavyweight computing companies such as Apple, Google and Microsoft.

At the physical layer, software also continues to become more complex, as a multitude of air interface standards are supported within a common solution, with a software-based approach providing much greater flexibility as standards evolve, rather than fixing low-level functionality in silicon with the risk of changes being required later.

Design for low power is a key differentiating ingredient between mobile handset software and traditional desktop or server-based computing solutions. This applies just as much to the high-level application software as it does to the physical layer software, and remains an important differentiator for semiconductor manufacturers, software providers and handset designers. A systems-oriented view which optimizes all aspects of the design for low power remains critical.

In the next chapter, we move our attention away from the design of the components of a mobile handset, and step into the shoes of a handset design team, to consider the typical process required to convert a set of component technologies and design concepts into a finished product design.

6 Product design

6.1 Introduction

In this chapter, we move the focus away from core handset components, designed to be used in many different handset designs, to the key principles involved in designing a particular handset. Drawing from the best design practice of handset designers who have created commercially and globally successful iconic handsets, two key themes emerge:

applying a *holistic* design approach;
embedding *desirability* at the start of the design process.

Holistic design is the integration and blending together of all of the individual design aspects – such as industrial, mechanical, user interface, hardware and software design – in such a way that a balance and synergy is achieved between the different design disciplines. The resulting handset should feel like a complete entity in and of itself, rather than a collection of things "stuck together."

Desirability is the expression of the needs and desires of the mobile user, encapsulated within the design.

In order to achieve a successful handset design, it is critical to map out the user's expectations, desires and needs at the beginning of the handset design. Elements of desirability must be injected into the overall handset design along each path of the design flow. This involves everything from choosing the colors and materials of the handset through to the way a user's finger feels a feedback click on the touchscreen and even to the ambience felt when a customer enters a handset showroom. Mapping this out at the beginning of the design process is important in order to ensure a consistency and adherence to the user expectations and desires throughout the whole design process.

Figure 6.1 provides an illustration of a typical set of stages which designers use when creating a mobile handset. These stages involve a wide range of disciplines including: trend, material, color, industrial, user interaction, mechanical, software and hardware engineering. As well as designers, the team needs to include marketing, manufacturing and procurement professionals. The key factors for achieving a holistic design through this design process are holding fast to the elements of desirability, openly communicating the vision across all departments, and working closely with the many different suppliers and partners to maintain authenticity in the original vision. This may sound difficult and time consuming, and it can be, yet, if the culture exists or can be created around delivering the best desirable and usable experience possible, then one is most of the way there. Understanding the key drivers of good handset design and the interplay of the main components are critical to defining and sticking to a good process of design.

We will now take a more detailed look at each of the design sections listed in Figure 6.1, in order to describe the key activities and best design practices. Additionally, we will describe the most important linkages between the designers and their internal or external customers. This allows the design process to be used as a strategic tool by marketers for positioning their handsets successfully in the marketplace.

6.2 The design process

Figure 6.2 shows a typical design process flow, indicating when all the individual design disciplines start and finish. The overall time from branding analysis to final manufactured handsets can take between six to eighteen months depending on the complexity of the handset design and how readily available the required components are.

Hardware and software design typically follow the mechanical and user interface design process, once the hardware platform and operating system have been selected.

One of the key deliverables of the mechanical design process is a demonstrable prototype which conveys the core design thinking. During this stage, the prototype can also simulate the proposed user interface

(1) Brand DNA Design
Mobile brand decoding
Design language
Product differentiation and segmentation
Brand integrity positioning
Value proposition and product price positioning

(2) Visual and Desirable Design
Global and cultural trends
Visual trends and feelings
Colors and materials
Treatments and finishes

(3) User Experience Design
Graphic user interface design
Usability scenario analysis
User interface platform design

(4) Industrial Design
Concept design
Product design
Product portfolio design

(5) Mechanical Design Engineering
Concept and product design
Prototype design
Appearance models

(6) Hardware Design Engineering
Industrial design and mechanical engineering iterations
Hardware platform design
Prototype product design

(7) Software Platform Design
Operating system partnership and platform design
Personalized services, widgets and content

(8) Manufacturing Production
Sourcing
Soft and hard tooling
Full production manufacturing
Approval testing (including testing and qualification)

(9) Channel Marketing Design
Channel design and management
Offline/online marketing and communications

Figure 6.1. Mobile handset design stages.

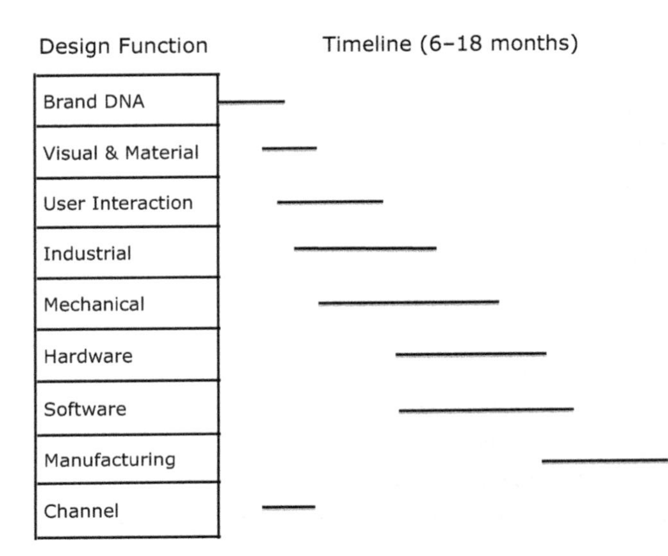

Figure 6.2. Mobile handset design flow.

and overall look and feel of the mobile handset. Based on design reviews and potential customer feedback, the opportunity is provided for last minute changes and customization before mass production of the mobile handset.

Prototypes of the industrial design, usually presented as 2D or 3D sketches or renderings, are created and reviewed by the design team. Once 3D renderings are approved, they are passed to the mechanical design team. Some organizations go further and first create space models, which are life-size 3D models of the design, allowing the look and feel of the design to be validated further. The mechanical team take the industrial design output, which is still somewhat conceptual, and design in detail all the physical aspects of the product, such as caseworks and battery housing. The mechanical design team need to work very closely with the hardware team to ensure that the PCB and all of its components can be accommodated within the design of the housing. A key output of the mechanical design process is the prototype mechanical design. This prototype is manufactured using soft tooling – a tooling approach to support the manufacture of the mechanical components for prototyping use, which is much cheaper than tooling used for mass production, but only supports a limited number of uses. The prototype allows the hardware and mechanical components to be assembled and tested by the design team. At this stage, the overall prototype can also simulate the proposed user interface and allow the look and feel of the mobile handset to be validated further.

Based on design reviews and potential customer feedback, this provides the opportunity for last minute changes in design aspects such as color, branding or customization ahead of sign-off of the design. The design is then transferred to the manufacturer for creation of the hard-tooled molds which will be used in the manufacture of the mechanical components for high-volume handset production. Increasingly, the relationship between the mechanical design team and the engineers associated with the manufacturing process is becoming more critical, due to the shortening product development schedules of modern handset creation.

Certain key platform decisions, such as the choice of hardware reference platform and software operating system, must be taken early, as otherwise activities such as hardware and software design cannot sensibly proceed. With these key platform decisions made, certain design disciplines may start earlier than others. For example, with the hardware platform and operating system chosen, the software design may start ahead of the mechanical design. This is possible because the software designers can start designing new applications if the screen resolution is known, on the assumption that any change in form-factor will not invalidate their design.

6.3 Planning a handset design program

In this section we outline the key best practice design principles which help to manage the design process more effectively and reduce design errors.

In planning a handset design program, all of the high-priority drivers for the project schedule need to be identified. These might include operator roadmaps, seasonal retail channel sales cycles and competitive windows of opportunity for new handset launches. For example, Carphone

Warehouse, a leading handset retailer in Europe, has seen an increase of more than 13% in handset sales during the Christmas season. Handset product planners must therefore ensure that good supply of their product is available at the beginning of this three month period to maximize sales potential. Having a phone available to meet the Christmas sales cycle typically means having the phone ready for volume manufacture by the end of August or early September. With the key delivery milestone defined for the whole development, project planners are then able to work backwards from this date to create the overall project plan to meet the optimal market window.

A key influence on the project plan is the potential for long lead times on certain manufactured items, such as customized displays or power transistors. Timely selection and procurement of long lead items is therefore critical to avoid the risk of delay to the overall project schedule.

A number of different companies may be involved in the design and manufacture of a handset design. For example, we have used "design team" to mean a collection of designers designing a handset for a client. This client could be an internal client within the company, where the design team is an internal department; alternatively, the design team could be a design agency working on behalf of an external company client. Further, when the designs are ready to hand over to manufacturing production, the manufacturing may be done in-house by the brand who is known as the original equipment manufacturer (OEM). Alternatively, manufacturing may be outsourced to a manufacturing partner known as an original design manufacturer (ODM).

Two major factors in good design practice are managing the interdependencies between each design discipline, and ensuring good and consistent communication between the different design divisions. In reality, what happens is that shortcuts or delays in some aspect of the program can take place due to the sudden need to launch the product more rapidly. This could be to meet a changing time window of market opportunity, a product release by a competitor, or the need for a sudden price change which changes design assumptions. Additional challenges exist when much of the design and the manufacture is outsourced to companies in very different time zones. For instance, Western Tier 1 handset OEMs who outsource to Far Eastern ODMs may face communication translation challenges, especially when the industrial division may be based in one part of the world and the manufacturing division in another. Good project management and regular communication during project update meetings between departments is essential.

Industrial, material, software or hardware designers all need a consensus on what are the key differentiating factors and experiences the handset must deliver. For example, in designing the best mobile TV experience, the form-factor, display, chipset, power, usability, services and content and antenna all need to work well collectively in order to deliver a great result. A good industry example is Research In Motion (RIM) with their range of BlackBerry devices. RIM have consistently ensured that their design team are focused on the most intuitive and secure e-mailing experience. As another example, the manufacturer INQ has created a number of mobile devices that focus on the best social networking experience.

6.3.1 Building a great design team

There are many factors involved in building a great design team. In the following, we list five key points that we believe are particularly important.

6.3.1.1 Design to manufacture

It is essential to have a mix of experienced designers who have a track record of having designed handsets which have successfully been launched into the market. This is particularly relevant when designing the user interface and embedded applications around specific mobile operating systems, as these can be very complex and time consuming unless you have good experience within the team.

The location for manufacturing can be important, not only for cost, quality and scalability, but also from a local sourcing point of view. Using good local skills could enhance the overall marketing brand story, for example "made in China" and "made in Switzerland" offer different expectations, yet both are highly capable and very professional.

6.3.1.2 Strong co-working between mechanical and hardware teams

It is important to have mechanical and hardware designers working very closely together. A good example of close integration was the development of the iconic Motorola Razr, which, in its time, was one of the thinnest clamshell handsets in the market. New keypad materials and mechanical structural designs had to be implemented for the first time, and continual negotiation with the underlying hardware platform designers was required in order to ensure that every single space-saving method employed around the hardware could be implemented.

With increased availability of modular, scalable, well-integrated hardware and software platforms, key points of visible differentiation on handsets are the external mechanics, form-factor and user interaction. This is especially true if the design is a touchscreen-enabled handset, as scope for product differentiation is limited by the need for a large screen. The combination of good mechanical and hardware engineering is vital – for example, a change in the overall physical volume of the handset may affect the hardware design significantly. In addition, if the design uses metal as part of the outer mechanics, there are possible impacts on the antenna design, so it is critical to review these aspects of the design carefully through regular design review team meetings.

6.3.1.3 Team communication

Project risk can be reduced by encouraging "overcommunication" – i.e. communicate early and often. A common process for identifying and managing risk is called *phase gate analysis*. The project is split into a number of clearly defined phases, with a review "gate" at the end of each phase. At the end of a phase, the team review the outcomes of all design activities, assess areas of improvement and potential risk, and decide how to mitigate any risks which have been identified. It is important to be extra vigilant at the beginning of the program. Continual dialog between internal and external clients on expectations is critical, including specifying the required quality of deliverables, providing approvals to move to a next phase, and keeping documented proof of agreed actions.

There is much experience in the industry of not getting this right at the beginning. The result is that the project will take longer, be more expensive, and have a different result from that outlined in the original brief. Nokia have historically had a very strong reputation for high-quality handsets, and we believe that this is due in part to the very strict discipline used in identifying and managing risk internally, but also through strict qualification and selection of their suppliers and partners.

6.3.1.4 Project management

We believe it is important to have someone manage the overall program who is extremely assertive and tough in dealing with high-pressure time-related matters. Excellent communication with all stakeholders and tough discipline are required. The program manager must be able to manage and balance timescales, resources, budgets and technical challenges. In addition, it is important to be flexible and be able to deal with schedule changes if market conditions change whence careful trade-offs need to be made to achieve a faster time to market. Holding face-to-face meetings at the manufacturing site is vital when final designs are handed over. An overreliance on communication via conference calls at critical project milestones is to be avoided.

6.3.1.5 Final handset price and time to market drivers

It is very important to respect the ongoing and changing market conditions that a handset design will be positioned for, without letting this be a constraint on the creativity of the designers. Typically, the final price of the handset, and hence the bill of materials (BoM), is pre-determined by the handset product portfolio managers or advanced procurement teams. From a marketing and sales perspective, this is where profits or losses are made. Designers need to respect the limits given to them, yet also be creative and innovative enough to push and test the boundaries. For example, a phone with a low BoM requirement does not have to look cheap. Sometimes "less is more": a minimalist design with simple usability can look and feel very valuable depending on the design language and product personality being designed for. Designers need to work very closely with the suppliers of new technologies, materials and platforms.

An example of how a low-cost handset design can be attractive for the mass market is the Nokia 1100, which was designed for developing countries and launched in late 2003. The handset has a keypad and front face which is dustproof, and the sides are non-slip for humid weather. The handset looks very simple and is very easy to use. The Nokia 1100 has been one of the best selling mobile handsets ever, with more than 250 million handset sales recorded.

In the following sections, we will further explain the key design disciplines shown in Figures 6.1 and 6.2. We will start with the core values and identity of the handset, referred to as the *DNA of the brand.*

6.4 Brand DNA design

The importance of the brand image and the brand story of the handset cannot be emphasized enough. An increasing number of lifestyle brands are entering the mobile market, including well-known names such as Dior, Tag Heuer, Porsche, Levi's and Armani. The essence of the brand, what it stands for and how it communicates its brand values all need to be embedded in the design of the mobile handset. It could be as simple as the color or even crystal materials used on the handset, or it could be the ease of application downloads and iconic design, such as for the Apple iPhone. As in the world of fashion, personalization and differentiation are required, but these come at a cost. One effective way of brand design association is through personalized branded applications and services which can be embedded in the handset or downloaded onto the handset over the air.

At the beginning of the overall design of the handset, it is important to understand what the brand on the handset stands for and how to maintain its integrity throughout the value chain from concept design all the way through to retail sales and after-care service. The author, Abhi, will never forget the day he walked into a large supermarket chain in the UK to see a broken and scratched LG Prada handset model sitting on the shelf with other handset models for sale, free of charge with a carrier contract. For Abhi, this shattered his perception of Prada products being of extreme high value, exclusive and very well crafted.

Typically, the desirability and brand DNA design starts first, feeding into the industrial, mechanical and user experience design, which can then all start in parallel. The brand DNA, or blueprint of the brand, is essentially a code of instructions which describe what the brand stands for in terms of values, aspirations, story and image. An example of this is how a US company called Monster created a range of headphones incorporating famous musical celebrity personalities into their products. They promote their core message as "putting the full musical experience into the headsets." One of the celebrities they worked with is Lady Gaga, and the headset they created around her is described as "holistically designed to deliver life's soundtrack as well as satisfy the passion for fashion." This expresses Monster's brand DNA and is reflected in the industrial design, the sound quality, packaging, innovative mechanical design, which pushes the boundaries, and the edgy channels to market in which they have chosen to sell the headsets. One example of where the brand DNA has not been carried through particularly successfully, from design to channel marketing with a strong consumer demand, is a mobile handset produced by Bang & Olufsen. Bang & Olufsen are known as one of the leaders in luxury stylish audio equipment for consumers. They decided to extend their brand and create a mobile phone with Samsung, which was launched in 2005. Their brand DNA is based on superior audio experience, intuitive usability, elegant design and an exclusive and personalized consumer purchasing experience through selective channels. However, consumers had difficulty in relating to these core brand values because of the difficult-to-use user interface of the handset and the limited distribution channels. Bang & Olufsen decided to pull out of the handset market in 2008.

Another example of how brand values can be incorporated in the mobile handset design upfront is the successful launch of the +YvesBehar phone in March 2011, by the Danish company Aesir Copenhagen. This contemporary high-end phone was designed by Yves Béhar and involved some of the leading mechanical, hardware and manufacturing designers in the world. During the initial technical design and project management meeting, the visionary CEO and founder of Aesir Copenhagen, Thomas Møller Jensen, made the brand values and DNA very clear through the design, development and manufacturing process. The brand DNA values of open collaboration, "Copenhagen guts" and its provocative feelgood and pure function were regularly used when making decisions on the choice of materials, technology partners, manufacturing location and even the packaging. This level of attention to detail and design ensured a very successful product launch.

6.5 Visual and desirable design

The aim of visual design is to build the appearance of the handset that will be desired by the target group. The outputs of the visual design could appear on a wallchart showing cutouts of photographs displaying architecture, fashion models, textures in glass objects, interior designs, fabric prints, geographical landscapes, cars, etc. This allows particular trends to be identified, which may lead to segmentations within the target user group. This display of information on a board is usually referred to as a "mood board."

Specifically, the following are typically explored in a mood board.

Product feeling: the environment the product will be seen in and the key messages that the product should communicate. Physical, industrial and user interface design guides, for example, considering form-factors or touchscreen gestures.

Color and material feeling: material design guidelines and color design variants such as different tones of shade.

Color material design: application of color and material feeling to the design concepts from which simulation prototype models are created.

These simulation prototype models are usually referred to as "mock-ups" or "appearance models," and are examples of the handset without any electronics or software inside. The purpose of such models is to convey at first sight and touch the initial design thinking, without incurring the cost and time of fully developing the handsets. These models are usually made in workshops with sample colors and materials, and can look exactly like the real finished handsets from the outside. These are very useful for handset sales people to demonstrate to potential buyers from major retail channels. Initial purchasing feedback can be obtained on factors such as the correct thickness, color, style, and even weight of the complete handset. Heavy material can be inserted into the handset to simulate the distribution of weight which the battery and electronics will exhibit when present in the final handset.

Visual designers keep a wide collection of the latest materials and colors that can be used on mobile handsets, and, just like the mechanical engineers, are very aware of the latest molding and final finishing techniques. A few examples of how materials from the non-handset world are crossing over to the handset world include leather, ceramics and carbon fiber. Increasingly, "eco friendly" and sustainable materials are being used, and further thought is being given to how the handsets are manufactured, packaged, retailed and disposed of in order to ensure that the brand message of "green" is authentic and genuine.

6.6 User experience design

User experience design for a handset can be considered as three main activities: the graphical user interface (GUI) design; usability scenario analysis; and user interface (UI) platform design. We will briefly describe each of these below.

6.6.1 Graphical user interface

This covers everything that the user sees, and sometimes feels, around the display area. It involves designing the navigation logic, the menu systems and the feature selection. Increasingly, this aspect of the mobile handset is becoming much busier as more applications are built into the typical smartphone. Some designs utilize a multiple home screen approach, allowing other screen displays to be revealed to the left or the right, by swiping with the finger. This provides additional flexibility, and a greater effective screen area, to rearrange the icons or apps that a user would like to access first or more often.

6.6.2 Usability scenario analysis

This is a storyboard of how someone uses the mobile handset in a typical scenario, such as sitting at an airport and using their mobile handset to play games, or to update their Facebook status with a photo they have just taken.

6.6.3 User interface platform

This area represents all points of the handset that a user interacts with when communicating with their handset – such as touch, audio input, audio output and sight. Some examples of user interface features which provide additional interaction points with the user include extra physical buttons to support taking photos or zooming the lens of the inbuilt camera from the side of the handset, or fingerprint recognition through swiping of the finger on a touch panel.

We provide a design phase list for user experience design, which provides a further sense of the key steps involved in a typical design process, in the appendix.

The user interface on the screen and around the keypad area is paramount for success in delivering an intuitive and simple user experience. This is driven by improvements in data bandwidth, the availability of rich multimedia capabilities and the availability of large numbers of apps on handsets. Various technologies are making it easier to navigate the user interface, although there is still much scope for further innovation. Examples include the widespread adoption of the touchscreen, adaptive user interfaces which change with the mode or context of use, voice input and multiple sensory outputs, including haptic (vibration-like sensation) feedback. A new type of vibration technology based on pulsating actuators is gradually making its way into gaming console accessories, and this technology could also be incorporated into mobile handsets. The promise is that this technology could re-create the feel of mechanical buttons on touchscreens. If this is the case, and the technology is sufficiently mature and priced low enough, this could start to appear in mobile handsets and tablets as well as other touchscreen-enabled consumer devices.

Getting a good sense of the usability needs early on in a project is important. Tools used to help achieve this include user journey storyboards, which detail usage scenarios and user journeys, and the development of target user segmentations. For example, fast text entry may be highly relevant for an 18 year old in Tokyo, whereas a display screen showing large images and clearer speech output may be more relevant for a senior citizen in New York.

Designing the complete user interface of a handset can take as long as six months, depending on the operating system and level of coding required. A good user experience is critical to achieve adoption of service use via the handset, both for operators and application developers. The Apple iPhone is a great example of how easy it is to download a large number of applications that can suit a wide range of users. Apple have demonstrated that even young children can download apps onto their mobile handsets in a very simple manner. By contrast, recently there have been several iPhone clones coming out of Asia for which the user interface design has not been well considered, and the result is a much less intuitive user experience.

One exciting area for user interaction designers in the future is to work closely with content and application providers to have profiling and pre-planning software integrated into the user interface. For example, such software could allow a user to wake up in the morning, switch on their handset and receive suggestions as to which is the best mode of transport to take that day for a family trip to the city center, the congested routes to avoid, family discount promotional meals and real-time alerts as their plans change. We cover this topic further in Chapter 7.

Increasingly, user interaction designers are including retail- and channel-focused applications on their front screen, which potential consumers can easily recognize and experience at the point of purchase. For example, the HTC HD7 Microsoft Windows phone displays T-Mobile TV, Netflix and Xbox Live on the front screen of their user interface. A trend in advertising of handsets is to show the handset screen displaying well-known social networking sites, music and movie downloads and mapping information. These applications are appearing on very attractive displays, in some cases 3D user interface displays. An example of this is the range of mobile handsets from INQ, owned by Hutchison Whampoa. INQ promote their handsets as social networking tools, and adverts for their products use images of the handset screen highlighting the latest social networking apps.

6.7 Industrial design

Industrial design is where art and science meet. Industrial design is about giving meaning to the form and function of the mobile handset.

In the section which follows, we shall provide an example of the set of deliverables, and the typical timescales involved, in an industrial design project for a mobile handset. The term "client" can refer either to an internal customer, such as a design department within a mobile handset company, or to an external customer, such as a design agency acting on behalf of a mobile handset company.

6.7.1 Phase 1: design exploration, three weeks

During this phase, the client provides input to the design team for the design brief and the design language. For example, a corporate design rule book may already exist which provides input on the consistent use of colors, fonts and design style, major component layouts and the hardware platform, if already available. A good example of the application of this is Apple, whose products such as the iTouch, iPhone and iPad all follow a consistent design language of simple, crisp, clean and thin with aluminum type casing or buttons.

Industrial design outputs include:

- design directions based on trend research, colors, materials and final finishing research;
- concept and idea sketches;
- 2D and 3D renderings, which are images of the shape and style of the mobile handset;
- initial design support from the mechanical design engineering team.

6.7.2 Phase 2 : design refinement, three weeks

During phase 2, the client provides feedback on the design concepts presented to them with their preferred choices. In addition, the client provides input into the creation of a plan for sign-off of each of the subsequent design milestones.

Industrial design outputs include:

- refined concepts from phase 1;
- fully featured 3D files;
- high-quality renderings;
- full color material finish definition;
- high-class appearance (simulation) model.

Phase 3 is the concluding phase, with the creation of the final appearance (simulation) model, which takes up to three weeks.

Timescales for deliverables of mock-up handset models vary according to where they are being designed and the size of the company involved. For instance, in China and Korea, the design times can be as fast as one week, at a cost of less than $US1000. In Europe and the USA, by contrast, design times could be up to three to four weeks at a cost of up to $US500 000. The ability for Far Eastern companies to achieve faster and cheaper design cycles is due to a mixture of lower wage costs, more human resources and a greater business emphasis on time to market whilst retaining the quality of the models.

The characteristics of a strong industrial design team include attention to detail, ability to communicate the product design story and a passionate and confident design spirit applied to the work they produce. Combining these characteristics will produce a consistent and recognizable "design language" with which the handset brand will be positively associated. For example, BlackBerry phones have always focused on the key handset experience of making it easy to write, send and read e-mails for business users. The iconic Qwerty keyboards found on their handsets have been used across several versions of their devices. Although initially designed for business users, teenagers are also buying BlackBerry devices as a fashion item. When RIM launched their first touchscreen handset, they created a product story around the handset being a business device as well as a multimedia and social networking device. Feedback arose within the press and media that this overloading of capabilities could be confusing, and that the touchscreen-enabled BlackBerry did not have a smooth user interface, compared with the Apple iPhone. In addition there was a limited number of apps available, which damaged the design story about leisure as well as business use. For this particular product, then, we would assert that there was not a consistent design-led product story and design language.

Within the mobile handset industry, there is an endless debate about convergence and divergence of handset device types. On the one hand, there is the "one device fits all" convergence argument, where one device is designed to be able to achieve a multitude of different functions. On the other hand, there is the divergence argument, where each handset is designed to harness one key application, such as having a traditional Qwerty keyboard to make writing e-mails very easy. A particular key trend in devices is the ability to access the Internet and the move towards cloud computing services to provide rich and powerful applications and services. This could either drive future handsets to become ever more powerful computers with a lot of on-board intelligence, or drive the move towards stripped-down, low-cost handsets with most of the processing intelligence in the cloud on remote servers. The industrial designer should be knowledgeable about the impact and trade-offs on form-factors, display size, usability, ergonomics and design language of major product and market choices such as this. In the coming years, we are likely to see an increase in the use of video calling between handsets due to the availability of faster mobile networks and WiFi hotspots. The acquisition of Skype by Microsoft and the partnership between Microsoft and Nokia could enable products with improved mobile video communication capability. This in turn will drive the need for clearer and larger displays and improved audio and camera capability. In addition, designs will need to provide greater ease of use for video record, play, edit and share modes.

The industrial designers and user interaction design teams need to work closely with the mobile chipset platform providers, who have good insights into future applications and services – and hence the size and power requirements of the chipsets. Examples include developments in mobile TV, GPS and near-field communication (NFC) applications.

Finally, good industrial designers ensure that they stay within the budget for the handset build cost, checking design decisions for cost implications such as complex machining. This is really important at the beginning of the mobile handset design process, in order to avoid delays and the risk of going over budget for the whole program later on. Since industrial designers have a good general understanding of all the major design disciplines, they are typically able to identify potential cost problem areas rapidly upfront.

6.8 Mechanical design engineering

Mechanical design engineers work closely with the industrial designers to refine the designs and construct mechanical solutions based on the selected design concepts. The mechanical designs are complete solutions that are suitable for high-volume manufacture. As well as the actual designs themselves, mechanical design engineers provide guidance on correct test support procedures, support for developing the process of building the handset, vetting of tooling suppliers and, as required, on-site support at the point of manufacturing. Mechanical design engineers accompany the design through the manufacturing process until mass production is underway.

Typical key deliverables and timescales for a mechanical engineering design process are shown in Table 6.1. The purpose of this flowchart is to demonstrate at which stages visual industrial designs flow into the mechanical design process. Additionally, the flowchart shows how central the choice of hardware components is to the whole mechanical design process.

Mechanical engineers are great problem solvers and have the ability to turn a challenge such as a slow or obtrusive hinge mechanism into a showcase personality for the handset. Using a state of the art designed mechanical solution can add to the overall look and feel of the handset. An example of a handset product where mechanical engineers worked extremely well with hardware and industrial designer engineers, and with component procurement, is the development of the Motorola Razr handset. The Razr, launched in 2004, broke the mold for thin and

Table 6.1. *Mechanical design engineering process*

Client input to Phase 1
Visual and material design brief
Industrial design concepts
Hardware basics

Phase 1: Mechanical design output (two weeks)
Mechanical feasibility study for concepts
Preliminary 3D engine and main components

Client input to Phase 2
Phase 1 approval
Mechanical requirements

Phase 2: Mechanical design output (three weeks)
Mechanical specification
3D engine and main components
Outer surfaces in 3D with manufacturability and compatibility with hardware confirmed
Preliminary component placement

Client input to Phase 3
Phase 2 approval
Key hardware components selected

Phase 3: Mechanical design output (five weeks)
Bill of materials (BoM) for mechanical design components
Risk analysis
Preliminary mechanical 3D design layout

Client input to Phase 4
Phase 3 approval
Technology choices selected
3D files and specifications of the hardware engine

Phase 4: Mechanical design output (five weeks)
Tolerance calculations
2D and 3D files
Test results

(*cont.*)

Table 6.1. (*Cont.*)

Client input to Phase 5
Phase 4 approval
Feedback analysis
Prototype parts analysis and definition
Phase 5: Mechanical design output (seven weeks)
Final 2D drawings and 3D files
Outer surfaces in 3D with manufacturability and hardware compatibility confirmed
Part split lines (visible parts)
Risk analysis
Review of 3D files with client, electronics designers and manufacturer for pre-tooling
Review of pre-tooling parts/product (steel or aluminum tools)
Measurement report check, visual check, test results review
Update and review of files and drawings for mass tooling with the client

elegantly designed handsets that looked unique when first launched. Most mechanical engineers stay very close to new and innovative mechanical component suppliers, especially locally sourced and low-volume suppliers who are willing to personalize and customize the individual finishing mechanical components or material. Increasingly, in the more expensive handset ranges, mechanical watch movement pieces are beginning to be used made of platinum, silver and gold – adding an extra level of finesse and brand statement to the handset. Even so, designers need to be aware of the challenges of antenna performance and adhesively bonding or mechanically fixing the exclusive material in such a way that a high-quality first impression is delivered.

Typically, the overall output from mechanical design engineering involves appearance models and prototypes based on the original industrial design and mechanical dimensions of the hardware engine platform. The use of these appearance models can provide valuable feedback when continuing to the next stage of design, as well as sales and marketing demos to potential channels to market.

6.9 Hardware design engineering

This phase of design engineering ensures that the hardware architecture meets the feature specification list of the handset requirements, including power management, electrical characteristics, mechanical volume and physical layout, and that it supports the appropriate software platform.

Whilst designing the hardware of the mobile handset, the designer and component procurement team work closely together to check that the original bill of materials (BoM) target price can be met and that the component lead times are manageable. Since certain key mobile hardware components can have long lead times of two to three months, sometimes the designer needs to be flexible in their choice of component suppliers and usually has multiple suppliers in case of delays or the sudden "end of life" of certain components.

Since handset hardware component design has already been covered in Chapter 4, we refer here only to the design integration aspects of the reference design platform. Essentially, the reference design platform is all the parts of the mobile handset except the outer casing and battery. This includes all of the electronic components, including chipsets, the electronic circuit boards and electro-mechanical components such as the display, keypad or touchscreen and speakers.

Reference design platforms for mobile handsets need to be modular, configurable, scalable and easy to "plug and play" additional upgraded features and functionalities onto, in order to create multiple handsets for different segments, using the same underlying core platform. A good example of this is how Qualcomm – a leading chipset supplier for handsets – have a wide range of chipsets that can be used in multi-platform engines, not just for smartphones, but also for lower-cost feature phones, as well as tablets, e.g. from 2G, 2.5G, 3G through to 4G.

In hardware design engineering, a strong area of focus is the balance between the technical parameters of available processing power, graphics performance and overall low-power management of the handset, and the form-factor parameters of compactness and thinness. The exact trade-off of these factors will vary depending on the segment the handset is positioned for – for example, a high-end smartphone or an ultra-low-cost handset for emerging markets. To take this thought further, note that some smartphones are increasingly being used for gaming due to their high-quality display, 3D graphics performance and very good sound systems all built in as standard into the design. All of this added functionality consumes a lot of power, and so accessory devices are appearing in the market which have inbuilt additional power and are able to house the smartphone in the middle of the accessory cover casing.

Advancements in cloud computing, allowing the bulk of the data to be held and processing to be performed on servers, are enabling new services for mobile and tablet users such as the iCloud service from Apple launched in 2011. Cloud services have the potential to allow typical applications that one may find on smartphones, such as gaming or uploading photos on mobile social networking sites, to appear in lower-cost feature phones, with service provision from the cloud.

Traditionally, suppliers of hardware reference design platforms have been reliant on their end customers, namely the handset manufacturers or even carriers, to inform them of their future requirements and roadmaps. However, an emerging theme is for platform vendors to work directly with lifestyle future trend analysts to help inform their chipset technology roadmaps. The purpose of this is to develop roadmaps which look beyond the shorter-term requirements posed by carriers and operators and to focus more on emergent consumer and lifestyle trends. Examples of this are bundling Bluetooth and GPS functionality on the same chip, and dual-core processor smartphones for increased mobile video and real-time entertainment purposes, such as gaming and real-time live location mapping.

6.10 Software platform design

Chapter 5 has covered, in some detail, software component design. In this section, we consider the design and integration aspects of software platforms and the enabling applications that can enrich the overall design of the handset.

The choice of software platforms, such as the mobile OS, is increasingly becoming a key driver in the choice of handset to be designed. This is because of the availability of application stores, richer user interfaces (including 3D user interfaces) and the availability of capabilities such as mobile Internet search. The drivers behind these trends were discussed in Chapter 2.

Software design engineers are becoming much more involved, and at an earlier stage, in the overall handset design process than historically was the case. A key reason for this is that the choice of the software operating system (Android, Microsoft, Symbian, etc.) has become ever more important. Software design engineers need to understand fully the expected user journey, the applications and services contained in the design brief, and must work closely with the marketing and user interface/usability teams in order to make sure the correct choice of operating system will allow the functionality specified. The Apple iOS operating system is used exclusively by Apple to build their own ranges of products. However, the Android operating system is much more open, allowing, for example, HTC to create multiple home screen user interfaces with which users can choose to personalize their handsets. This approach also allows multiple handset companies to personalize their handset user interfaces to some degree, providing some level of differentiation between each of the handset companies which use the Android operating system.

The availability of these mobile operating systems and other software, such as on-device shopping portals, presents challenges for developers of innovative applications and services. Which platform should they design for first? How will they get high-visibility and high-volume downloads? Will the shared revenue business models be lucrative enough? In addition, there exist an increasing number of partnerships between carriers and content providers, which creates a busy landscape for software design engineers to innovate within. Very good opportunities exist for the best to stand out from the crowd and design something unique for selected markets. The rapid adoption of advertising-funded music channels and voice over Internet protocols (VoIP) on mobile handsets are exciting ways for software, hardware and user interface engineers to work creatively with the marketing value proposition teams to ensure they can deliver a consistent service. An example of how this approach has manifested into deeper relationships between some operating system providers and service providers is the acquisition of Skype by Microsoft.

Consolidation and partnerships between handset OEMs and software technology providers are likely to increase due to the growing role of an open software platform in the handset design. Examples of such partnerships and acquisitions are Nokia's partnership with Microsoft, the acquisition of a leading user interface platform provider TAT by Research In Motion, and Google's acquisition of Motorola's mobile handset division.

6.11 Manufacturing production

Although the main focus of this book is on handset design and not on manufacturing, "designing for manufacturability" is of paramount importance. Without consideration of manufacturability, there is a significant risk of ending up with "blue sky thinking" and interesting design concepts, yet no product in the marketplace – because the design could not be manufactured correctly.

In 2012, the majority of handsets sold globally are manufactured in Asia, with a large share being manufactured by Chinese ODMs. According to HIS iSuppli, a leading research firm, four out of five mobile PCs are produced in Shanghai, and 25% of all mobile handsets are manufactured in Shenzhen in China. The quality, speed of response and flexibility of the ODMs is extremely high. Yet, as for any subcontractor, designers need to ensure strong communication channels are in place to provide consistent adherence to the industrial and mechanical design handoffs to manufacturing production. Some ODMs, in order to maximize profits, seek to keep the manufacturing and production lines fixed on standard product designs rather than to deviate to full custom designs. Continual negotiation and education are necessary between designers and manufacturers, so that each party is aware of what is possible, when new ways of manufacturing need to be considered (and at what cost), and when it is appropriate that the handset designs are slightly customized or deviate from the norm. One example of a very innovative materials company that works closely with designers and handset manufacturers is Trexta, based in Turkey. Trexta specializes in plastic, fabrics and leather, and it has created state of the art unique manufacturing techniques in leather production for some of the most prestigious and expensive mobile handsets in the world, as well as some of the leading iconic brand handsets.

6.12 Testing and qualification

Testing and qualification details and processes are outside the scope of this book, although the importance of thorough testing and the acquisition of correct qualification approvals need to be understood and appreciated by professional manufacturing production engineers.

When designing the handset, careful consideration needs to be given to which radio standards will be used, e.g. GSM, CDMA, 2G, 2.5G, 3G, 4G, and how many bands the handset will operate at, and in which countries the handsets will be sold. The choice of standards, radio bandwidth and number of bands will impact the type of testing and qualification for carrier approval in terms of time and cost, and so this must be taken into account in the overall handset design process. Testing and qualification of the handset is critical. An example is the launch of the Apple iPhone 4 in June 2010. Soon after its launch, news started to appear that some users were experiencing signal loss, depending on how the handset was held. This led to some press commenting that the antenna design in the mobile handset may have been the cause of the problem. Having manufacturing, hardware and mechanical designers along with test engineers working very closely together can significantly remove risk from the test and qualification process.

6.13 Channel marketing design

An area of handset design which is often overlooked by many handset OEMs and retail channels is the ability to use design to influence the consumer at the point of sale. Examples include the opportunity to create eye-catching demonstrations on live handsets, personalized packaging and clear and easy to read information on why this handset is different to the one next to it on the shelf.

In 2008, an auction for a custom-made Bugatti high-performance sports car in California took place. One of the interesting value propositions was for the person who bought the beautiful and elegant Bugatti to fly across to the factory and meet the designers and tell them what features they would like. Eventually the Bugatti sold for $US2.9 million. Can these same dynamics of personalized channel marketing be applied to a $100 handset, where the user designs their own handset online and through social networking and group purchasing goes on to purchase the product? Some innovative companies are already working on this challenging idea.

In 2009, Sony Ericsson launched the GreenHeart™ initiative aimed at producing an environmentally friendly range of mobile handsets including eco-friendly design, packaging, production, shipping and recycling. No paper manuals are supplied, and even the charger is designed to use less energy. The Sony Ericsson Aspen, launched in 2010 as part of the GreenHeart initiative, is constructed of recycled plastics and water-based paint and is marketed as a "phone that helps you reduce impact on the environment." The Aspen is a good example of how Sony Ericsson has applied its core brand values of sustainability throughout the whole design process, from handset design to disposal, and has communicated this at the point of retail as a strong differentiating value proposition.

The following case study demonstrates how the complete design process is influenced by the introduction of new innovative technology. Capacitive touchscreens have now received mass adoption in smartphones and some feature phones, but not so long ago they were very disruptive to the traditional design process.

6.14 Case study: capacitive touchscreens in mobile handsets

In this case study, we will take you through the adoption of capacitive touchscreens into the mobile handset and discuss how handset designers played a key role in working collaboratively with the whole mobile ecosystem supply chain.

Capacitive touchscreens are found in Apple iPhones, BlackBerry Bold and smartphone handsets from Sony Ericsson, Nokia, Samsung, LG and

HTC. According to Display Search, a market leading display research company, 40% of mobile phones are likely to have touchscreen interfaces by 2015.

Capacitive touchscreens use an insulator, such as glass, coated with indium tin oxide. The human finger acts as conductor, and when it touches the touchscreen, an electrostatic field distortion is created and measured as a change in capacitance, which is then translated into a signal by algorithms running on a chipset.

When capacitive touchscreens were first introduced to handset designers, innovators were seeking to replace resistive touchscreens, which were used in products such as the early Palm Pilot PDAs (personal digital assistants) with a stylus. The advantages capacitive touchscreens have over resistive touchscreens include: improved display clarity (due to the lack of an air gap found in resistive touchscreens), improved wear and tear and the ability to support very intuitive finger input gestures with a fast response. The disavantages of using capacitive touchscreens, at the time, were that they were new to the market and more expensive than the competing resistive touchscreens.

When capacitive touchscreen technology was first presented to some of the top Tier 1 handset manufacturers in the world in 2001–2002, their initial reaction was that capacitive touchscreens would never be used in mobile handsets. Their main reasons for this opinion were that there was a lack of a clear value proposition of how the technology could be used, that the technology was perceived as immature and that there was no proof that customers would find the benefits attractive.

Capacitive touch technology presentations at that time were mainly given to heads of marketing and advance technology evaluators, as well as to research and development teams.

To seek to counter the main objections, the supply chain was critically re-examined, leading to the identification of network operators as potential key stakeholders (and therefore gate openers). The value proposition was re-worked to include usability results along with demonstration and evaluation units, which showed how content and applications could be accessed more easily using capacitive touchscreens. Key mobile operating system developers, display vendors and specific mobile chipset developers were engaged as partners to provide a more credible and less risky value proposition with proven and very knowledgeable handset partners. When the capacitive touchscreen solutions were presented again, members of the senior management teams started to attend the presentations from the handset and carrier companies. It was also noticeable that industrial and user interaction designers were beginning to express serious interest.

The industrial and user interaction designers had the vision and creative insight to see the value of capacitive touchscreens, and they contributed towards internal buy-in in order to evaluate capacitive touchscreens further. Eventually, it was with the help of mechanical engineers, with a final push from software engineers, that the breakthrough came. Feasibility studies were carried out that clearly showed that capacitive touchscreens could be manufactured at high quality and in high volumes.

What was not anticipated in the early days of capacitive touchscreen adoption in handsets was the need to enhance software operating systems to incorporate touch input. Considerable software development resource had to be used by the handset manufacturers along with their user interaction designers. In addition, hardware integration and factory testing of the touchscreen and display coupling were needed, as was hardware design integration of the touchscreen controller. In the meantime, the visual and material design engineers were busy analyzing how moisture and gloves affected touch input, whilst the brand DNA designers were creating and embedding the "touch story" throughout the design process. Channel designers had to design the right channel partners and best forms of communication to demonstrate the touch value proposition.

In 2007, LG Electronics partnered with Prada to launch the world's first fashion handset with full capacitive touchscreen. However, it was not until Apple launched their first ever iPhone handset, which featured a capacitive touchscreen, that the market really took off. Apple provided a solution with a highly intuitive and very accurate response to finger touch input combined with their very well planned channels to market and memorable packaging. Both LG Electronics and Apple had very close control and influence over their own mobile operating systems which were used in these mobile handsets. This provided them with the advantage of incorporating gesture-based navigation features such as pinching and swiping relatively easily.

In conclusion, the adoption of the capacitive touchscreen demonstrates how closely each design discipline (including industrial, mechanical, user interface, hardware and software divisions) needed to work together with a common goal of making a more intuitive and engaging handset design.

6.15 Using the design process as a strategic tool for innovation

Mobile handset designers sit right in the middle of the mobile ecosystem value chain, and increasingly act as a rich source of innovative ideas for new design techniques, materials and ways of engaging the user.

Figure 6.3 shows how some key players of the mobile ecosystem can influence and partner within the mobile design process. The shaded parts highlight those technologies and applications, design functions and enablers that we believe today are impacting the handset design the most. This does not take away from the value of the other technologies and design disciplines featured, rather it suggests that the center of gravity is in those highlighted areas; no doubt in time these will shift to other areas.

This figure purposely looks somewhat like a spider's web, demonstrating the interconnectedness required. For example, if a pico-projector company wanted to embed their miniature projector inside mobile handsets, they would need to work very closely with the brand DNA, visual, industrial, user interaction and software designers, as well as the hardware designers.

As boundaries merge within the mobile ecosystem, it is more important than ever to "think outside the box." It is important to explore how suppliers of technology and application developers can engage collaboratively with the design community to share their ideas of future trends in mobile handset design. Mobile handset companies are increasingly holding internal behind-closed-door innovation days, where they choose design houses and new emerging technology suppliers to showcase their vision for the future through concept designs and prototype

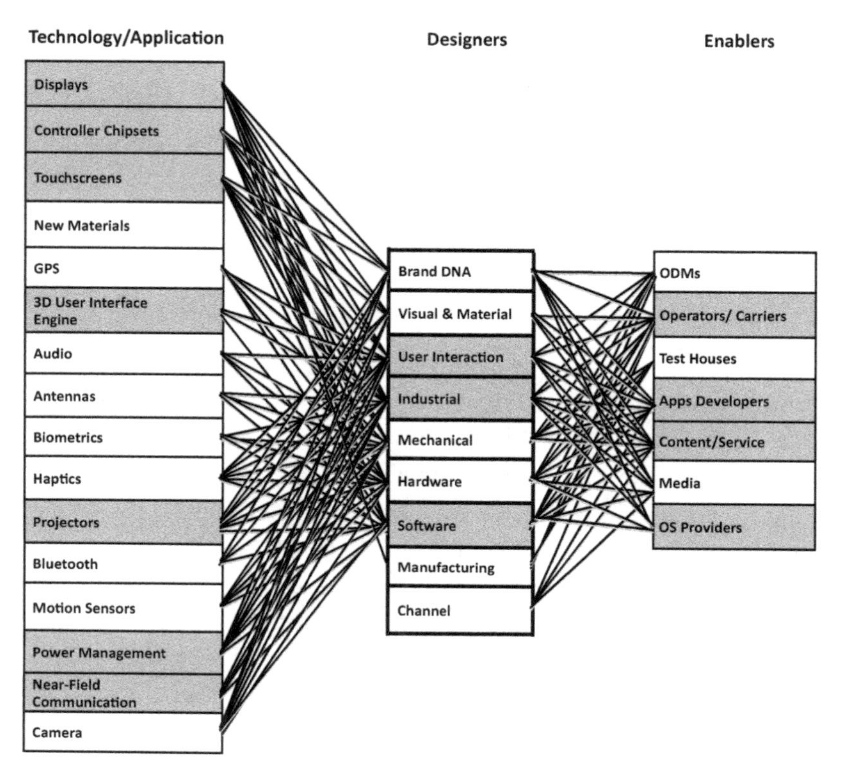

Figure 6.3. Suppliers to the mobile handset – strategic engagement and influence.

demonstrators. In some cases, if the demonstrations are very innovative and match the mobile handset company's aspirations, the technologies and designs will be fast-tracked into commercially available mobile handsets.

6.16 Inclusive design and accessibility

All of the design processes and disciplines covered so far can equally be applied to the design of mobile handsets for users with disabilities. Sadly, not enough attention has been given to date to the needs of these users by some of the major players in the mobile industry. Fortunately, this is changing due to factors such as new government legislation in some parts of the world, charities raising the awareness around the digital divide amongst users with disabilities, and entrepreneurial companies creating inclusively designed handsets that increase accessibility for all mobile users. It makes good design and commercial sense to design the handset so that it can be used by everyone no matter what their age, geographical location and physical abilities.

Areas of technology innovation which we expect to drive improvements in inclusively designed products include speech input and output user interfaces and use of haptic vibration feedback on touchscreen displays.

Currently, research is being carried out around the idea of making the mobile handset user interface more adaptable when a person with disabilities is moving from one room to another or outside the house. The availability of motion sensors and Bluetooth proximity sensing in outdoor locations is on the increase in major cities around the world. These systems can be used to relay dedicated emergency support or presence of facilities straight to the handsets. Although this is still in the early stages of adoption, the area of mobile health monitoring in the house is likely to increase as large companies such as Telefonica and Verizon move into this area. According to research conducted by Berg Insight Remote, health monitoring generated €7.6 billion ($US10.01 billion) globally in 2010.

From a handset content perspective, exciting applications are emerging allowing blind and partially sighted people to recognize colored objects, and even shapes and labels, using a server-based application and the camera on their mobile handsets.

6.17 Conclusion

Taking a holistic approach to handset design and working closely with suppliers of mobile technology and services is, in our experience, by far the best way to work to achieve compelling products. Keeping focused on the end user and keeping the brand DNA in mind throughout are critical in delivering a lasting and impactful user experience.

Encouraging designers to meet suppliers of handset technology and services routinely to discuss technology roadmaps, handset integration challenges and likely advancements in the technology is always beneficial to both parties.

It is becoming easier to make handsets with relatively low functionality, at lower costs and lower volumes. This will make it easier for new entrants to come into the world of mobility and "brand slap" their logos on the handsets. Will you ever see, next to the designer perfume stores in the supermarket, designer handsets? Perhaps one day the shelf space will be incredibly packed, and so each handset brand, service, cost point and design will have to stand their own ground.

The beauty of handset design is that it is international, and we have seen faces light up positively in design review meetings in Seoul, Tokyo, Helsinki, Stockholm, London, New Delhi and Chicago, as soon as new teams see design concepts and visual languages for the first time.

Striking a good balance between evolutionary and revolutionary technology and design practices is an exciting challenge, and one which will continue to create some magical and highly desirable mobile handsets in the future.

With the future in mind, in the final chapter, we provide a review of many of the emerging technologies and trends for handset design. These are the key issues which we believe are most likely to impact strongly the design and key components used in the mobile handset. By looking at these trends, we can also provide a view of how we, the end users, may be interacting with our mobile handsets in the coming years.

7 Future trends

7.1 Introduction

How the future will look in one, ten, or 100 years from now, no one really knows, as even how the mobile industry will change in three months' time can't always be predicted. All we can do is share trends and insights into what we believe the environment will be like in the future, then assess how these trends are likely to shape and influence mobile handset design.

In this final chapter, several possible environments and landscapes that may exist in the future are explored, and the key technology drivers that may influence the handset design are identified. In order to match the possible environment, trends and landscapes with potential design changes, we are going to tell you a story. This story is of a woman going through her daily life in a few years' time from now, using her handset to achieve everyday things. We will then discuss how the specific new technologies, applications, services or usage scenarios implied by the story fit into the typical handset design process, as described in Table 6.1 in Chapter 6. We repeat that sequence of events here, for convenience:

choice of materials,
user interface,
industrial design,
mechanical design,
hardware and software platform considerations,
channel to market,
design flow.

The idea here is that as future technologies, applications and services emerge that we have not detailed, the reader will be able to match them with the handset design flow principles described in this book and have an estimate of the likely impact on the design of the handset. Additionally, to ensure an active commentary on new and emergent trends and to allow us to make observations on future developments in the mobile handset industry, we have included details of a mobile handset design website created in partnership with the publisher of this book, which can be found at the end of this chapter in the summary section.

In selecting potential emerging technologies, key factors in the choice of technology center around how readily it can be integrated into a mobile handset and be manufactured at an appropriate price point within a reasonable development cycle and at relatively large unit volumes. We believe that this is the real test of all future handset design concepts. For example, it is great to see additional sensors, such as a rolling weight, being distributed and integrated into a demonstration or concept handset to reflect where one could navigate to when walking right or left as a pedestrian. However, one may question whether it can be manufactured at a price point and timescale that will appeal to the target users. It is in this area that industrial designers and product managers work closely together in order to bring reality to the designs.

One vision of the future which we think is particularly exciting is the opportunity for our phones to understand us better and become much more useful companions for a range of everyday activities. Justin Rattner, chief technology officer of Intel, captured this idea at an Intel developer forum:

"Future devices will constantly learn your habits, the way you go throughout your day. They'll understand your friends and how you're feeling. Maybe more importantly, they'll know where you're going and anticipate your needs."

7.2 The journey

Sarah awakes and gets ready for a productive day at the office. She drives to the railway station and parks her car, making the car park payment through her handset. Sarah walks past the ticket counter and waves her handset near the ticket machine. The ticket machine gives her a ticket at the best discount for that particular time of day and delivers an electronic copy of the *Financial Times* straight to her handset. As she sits on the train, she lays down her handset on the table and it automatically projects the display onto the back of the seat in front. A large keyboard is projected onto the table. She starts writing a report that needs to be submitted later that day, and then sees a message on her screen that a more in-depth article in the *Financial Times* could be relevant to her report. She orders the more detailed article through her handset. As she arrives at her final destination, she leaves the train and holds her handset in the air. She scans her environment, and, looking through the display, she finds the most suitable bookstore for her needs. This bookstore is offering recyclable paper folders and printing facilities, just what she needs in order to print out 20 copies of her report – at the press of a button from her handset. Her handset also shows her that included in the price of the printing she can get a 10% discount at the coffee shop next door as she waits for the print run to be ready.

After 20 minutes, with the reports in her hand, and listening to the latest music track from her favorite band downloaded onto her handset, she leaves the coffee shop and heads towards her office. As soon as she enters the office, her handset lets her know that three of her meeting attendees for her report presentation will be attending the meeting via video call. Her handset informs her that the presentation report has already been e-mailed to the remote attendees, and video conference facilities have been set up. The presentation and the rest of the morning meetings go well.

On the way to lunch Sarah is alerted by her handset that she has a group yoga session today. She decides to visit her local department store, where at the cosmetics desk she smells a beautiful calm and relaxing scent coming from a very fashionable bracelet-shaped handset. The lady at the stand explains that this handset can sense your stress levels, massage you to help you relax, and then produce a spray of perfume according to your mood. Sarah smiles when she discovers that a pair of these handsets can be bought and set up so that you can sense your partner's mood and send scents and vibration massages to each other. Sarah decides to purchase a pair of the bracelet handsets. She reaches into her handbag and gives the shop assistant her old handset. The assistant slips a mobile network carrier tag onto it, allowing Sarah's old handset to be recycled later. In return, she hands Sarah the pair of new handsets without any case or packaging. Sarah slips the new device around her wrist and walks back towards her office.

The route back is via a different path – it's the one her handset indicates is less polluted. The route has been calculated by tracking live pollution emissions through sensors in her handset as well as the handsets of people around her, with pollution patterns calculated through a cloud of networked servers. A cup of free herbal tea is already waiting for Sarah when she walks into a new café. This drink was pre-ordered – her bracelet handset automatically ordered the trial tea since it knew Sarah usually has a cup of tea before ordering her lunch.

The day passes quickly for Sarah, and she soon finds herself in the train heading back home. She notices a friend's stress levels suddenly rising as she looks at her bracelet handset, and is just about to send her a "text massage" when an alert flashes across the curved display. She had set her handset to detect anyone nearby who may be interested in supporting a local charity marathon she is running in, in three weeks' time. Her handset tells her that three seats away is a fellow runner, and so she engages him in a conversation about running tips, and they both share live 3D images and videos of the running track through their handsets.

At the end of her train journey, Sarah disembarks, enters her car and heads home. She lays the handset down on the dashboard resting area, where it is wirelessly recharged. She sees on the projected display that her presentation report proposal has been accepted and that a follow-on meeting has been scheduled in Delhi, for the following week.

On arriving home, Sarah is greeted by her children. The house temperature and lighting adjust automatically to make her feel at ease, based on information about her well-being sent from her handset. A documentary on India appears on the large-screen television – as well as details of the charity she will be running for and recommended running routes close to the proposed hotels in Delhi where she will likely stay. Later on, she puts her children to bed, and her daughter asks her for a bedtime story.

Sarah takes her bracelet handset off, unrolls a large flexible display with 3D animations, and reads the text from a children's story. As the children settle down, Sarah switches her handset to silent and tells them about her day.

During the rest of this chapter, we will unpick Sarah's journey into a number of different and promising trend areas, and explore these trends in more detail.

7.3 Designing for the environmental footprint

Environmentally friendly features and services have been designed into and around Sarah's handset by the handset manufacturers, retailers, network operators and third-party software developers. Specific examples include the use of recyclable materials in her handset, the presence of less or no packaging when acquiring her new handset, disposing of the existing handset when exchanged for a new one, and even detecting travel routes where there is less pollution.

Greenpeace, a non-governmental environmental organization, creates a guide to greener electronics, and ranks leading consumer electronics companies including mobile handset manufacturers on how "green" they are. Examples of specific measurement criteria are:

chemicals management;
polyvinyl chloride (PVC)-free and/or brominated flame retardant (BFR)-free model and phase-out timelines;
individual producer responsibility;
voluntary take-back;
amount of product recycled;
use of recycled plastic content;
global greenhouse gas emissions reduction support;
carbon footprint disclosure;
amounts of renewable energy used;
energy efficiency of new models.

As of November 2011, Nokia was the third top rated electronics company by Greenpeace behind Dell and Hewlett Packard. Further examples of handsets being developed around the environmental theme include a handset from Samsung called Evergreen. The Samsung Evergreen handset was sold by AT&T and was promoted as an eco-friendly handset. According to Omar Khan, chief strategy officer of Samsung Mobile, the handset is made of 70% post-consumer recycled plastics, and as a whole is 83% recyclable. It also features reduced packaging which is recyclable, a CD to replace the user manual and an Energy Star qualified charger. The area of power management and charging will be discussed further later in this chapter.

Other manufacturers that are producing eco-friendly handsets include Sony Ericsson, with their GreenHeart portfolio, launched in 2009, and Motorola with GRASP, which is a green handset featuring BFR- and PVC-free housing that is 100% recyclable.

With increasing awareness of environmental issues by the general public and governments, more attention is expected to be given to this area in designing environmentally friendly mobile handsets, as well as the issue of disposal of handsets.

In the UK, the network carrier O2 (part of the Telefonica Group) is hoping to create an industry standard with the launch of a new rating system that charts the carbon footprint of handsets on its website and in stores. However, the scheme only covers handsets supplied by O2 and did not include the iPhone from Apple at the time of writing this book. Additionally, O2 have been very active in recovering over 0.5 million handsets as part of a recycling initiative.

In June 2010, *The Guardian*, a UK newspaper, discussed the carbon footprint associated with the use of mobile handsets on its 'Guardian's Green Living Blog." Quoting from this blog:

> "The footprint of your mobile phone use is overwhelmingly determined by the simple question of how often you use it. One estimate for the emissions caused by manufacturing the phone itself is just 16 kilograms of carbon dioxide emissions, similar to the carbon footprint of about 1 kg of beef . . . if you are using your mobile for an hour each day, the total adds up to more than 1 tonne per year – the equivalent of flying from London to New York, one way, in economy class."

In future, more focus will be given to eco-friendly handset designs, choice of materials and manufacturing processes, as well as how one disposes of handsets.

7.4 The "smart" journey

During Sarah's journey, her handset was working smartly in sensing her needs and delivering the right user experience at the right time and place through the use of context and data mining.

In the future, our handsets will be used less for voice calls and more for interaction with data and services. This is already apparent as you see more people staring into their handsets rather than holding them against their ears. This raises again the issue of whether we should be calling the mobile handset a "handset" or even a "mobile phone," when in some cases people are using their mobile device on a table in speaker-phone format for a conference call, or using headsets and wireless eyewear rather than their hands. As wearable mobile technology becomes more mainstream in the future, it is believed that one-handed and two-handed use of the mobile device will still be the dominant method of interaction over the next few years at least. Voice calls will still be an integral part of the mobile device, yet they are already becoming buried within the content and service functionality, at least in the smartphone segment.

Forecasts by the research company Gartner Group suggest that 80% of mobile phones will be equipped for data communication by 2013.[1] We are in the midst of a very significant growth in the smartphone market, which in turn is driving the uptake of mobile data usage. This usage is very uneven, however. In September 2010, a senior Motorola executive estimated that just 5 to 8% of US mobile users were consuming 60 to 70% of the mobile data traffic. But he also noted that smartphones accounted for 45% of all US handset sales. Based on 2011 fourth-quarter earnings results from leading handset OEMs, Canaccord Genuity analyst Michael Walkley believes that global handset unit sales increased to

[1] Source: July 19, 2010, http://mobile.credit-suisse.com/index.cfm?fuseaction=mOpenArticle&coid=284831&aoid=286573&lang=en

455 million units in the fourth quarter of 2011, up 13% from 402 million units in the third quarter. Smartphone sales increased 34% sequentially to 159 million units. It is clear, we believe, that data-driven services have entered the mainstream, and will continue to be a source of innovation in the future.

Many opportunities exist to differentiate data-oriented handsets, for example by the design of relevant and engaging services and applications. Handsets which consume more data also consume more power. Significant opportunities exist to innovate across many aspects of a handset design to ensure that data is managed in a power-efficient manner. This may be achieved through improvements to the power efficiency of hardware components, or in software algorithms which manage the hardware more intelligently. In addition, further innovation opportunities exist to display the ever-increasing quantities of data available to the user in an intuitive user interface format. An example would be designing the user interface to make it easy to move between different modes of communication, such as Facebook to e-mail, and to receive text messages and Twitter updates through the same interface if required.

A big challenge for the network operators is ensuring that the network can support the increasing use of data from smartphones, whilst developing business models and data tariffs which can provide users with access to data at a reasonable price. In the UK, in December 2010, Orange, a major network operator, offered subsidies on the sale of Apple iPads to encourage customers to purchase these products. However, the promotion was not so well received. We believe this was due to the length of contract that a consumer had to sign up to and the lifetime cost of data access through the monthly usage tariff. Because there is a high cost to the operator of providing additional network infrastructure, and there is always a shortage of available spectrum, there is growing interest in other approaches to transporting data. Some approaches include the offload onto WiFi and the use of femtocells (small cellular base stations) to make better use of spectrum in a localized area.

Returning to Sarah's journey, we notice that in the background her handset is profiling her journey and anticipating what newspaper she likes to read and which coffee she prefers, and is automatically making permission-based payments relevant to her journey. This activity takes place based on her current context, for example whether she is walking outside, driving in her car, traveling by train, or is in the office or at home. Her handset is using the context of her current situation, and what is currently important or relevant to her, to access and filter data in the cloud. Services are then able to provide relevant and timely information for Sarah as well as make correct decisions on her behalf.

Cloud computing will greatly impact the handset of the future, allowing data to be delivered seamlessly whenever and wherever it is relevant to do so. The continued increases in processing power in handsets and the availability of higher data speed mobile broadband networks will encourage new service delivery mechanisms to access data when required from the cloud.

Juniper Research estimates that the market for cloud-based mobile applications will grow by 88% during the period 2009–2014.[2] From a base of $US400 million in 2010, the market could reach $US9.5 billion by 2014. As an example, Carphone Warehouse, a major mobile handset retailer in the UK, launched a cloud-based music service in August 2010. On offer is a music subscription service featuring music from major record labels at a cost of $US 47 per year with access to over six million songs available on iPhone, BlackBerry and Android handsets.

From a handset design perspective, we expect to see an increase in the range of display sizes and form-factors and more uptake of wearable handsets. A key success factor will be the ability to personalize a particular device for the type of service usage. This could include user interfaces which adapt as you change context, for example when you enter your house or get into your car. The availability of large quantities of information for the user to interact with and consume, drives a trend towards having larger screens and more tablet-size form-factors, typically with touchscreens. Having a larger screen is one approach to managing information, for example by providing multi-layered data presentation

[2] Source: February 22, 2010, http://marketingcharts.com/direct/cloud-based-mobile-market-to-grow-88-12043/

allowing viewing of your social networks and financial data at the same time as reading a book and listening to music.

As the user engages with a cloud-based service via their handset, there needs to be a consistent and intuitive approach to providing a seamless user experience. One of the challenges in the mobile handset industry today is: Who actually designs and owns the user experience? Is it the network operator, who provides the service; is it the branded handset OEM, who has a pre-installed apps store; or is it the branded operating system provider, such as Microsoft or Google? Whoever it is, or in whatever combination, the question is: Can they all work together?

Vodafone, a leading global network operator group, launched a cloud-based service in September 2009 called Vodafone 360. The service offered a consolidation of the address book, enabling users to link their phone contacts to social networking sites such as Facebook, Twitter, MSN and Google Talk. Vodafone worked with Samsung to develop two Vodafone branded handsets, customized to offer the Vodafone 360 service. Providing seamless synchronization services is technically challenging, and there was considerable press coverage of the difficulties some customers had in synchronizing e-mail accounts and social networking contacts between their Vodafone branded handsets and the 360 cloud-based back-up service.

The importance of good data synchronization has created opportunities for other companies. Microsoft has positioned its Windows Phone 7 operating system within a new range of handsets on the basis of providing a great experience when synchronizing contacts. As different industry players vie for the consumers' attention, significant handset design integration problems can emerge. Imagine, for example, trying to incorporate two services such as the Windows Phone 7 operating system and the Vodafone 360 service onto the same handset. Areas of potential problems can arise within the design; for example, where in the handset do you store music or games? Would it be Microsoft's "people hub" area or Vodafone's 360? Equally there are real challenges in synchronizing the address book and calendar and transferring contacts between different systems. Achieving a good user experience with a hybrid of different solutions has historically been very difficult to achieve.

We can expect there to be multiple operating system platforms in the future. In addition, it is already the case that one device may host multiple application stores and cloud-based services – for instance, from the platform provider and from the OEM, network operator or other service provider. Handset designers will need to choose very carefully which partners to align with first as new solutions emerge, to ensure a smooth path of design integration. We believe the advantage will go to those handset companies that either own or can control closely the complete vertical supply chain from content, apps, billing and operating system through to the OEM brand. Apple does this well today, though there will undoubtedly be others, as suggested by Amazon's launch of the Kindle Fire device and ecosystem, Nokia's alignment with Microsoft and perhaps realignment through Google's acquisition of Motorola. Questions remain around the role that the network operator will play in the future, with many moving into cloud-based services, leading to opportunities for distinctive niche brands.

7.5 The mobile home

As we continue to follow the journey of our character Sarah, we find that her home automation system has set the temperature controls to her mood and that her television has automatically selected channels that she would find interesting and relevant based on her day's activities. All of this information about Sarah would have been shared between Sarah's mobile handset and the smart appliances via a home network.

Presently, Microsoft, Google, Motorola and several carriers are offering mobile services and devices that can be used more effectively in the home environment. Users of these services need to see a consistency of service as they step into the home, for example synchronization of data and transparency regarding pricing. Data can be provided over the network and reside either in the cloud, on the device or on a local server in the home. Controlling how you access, update and share this data on the move, in the home or away from your home will be a key method of monetizing the user experience, and hence will be very attractive for the handset carriers, OEMs, app developers and OS providers.

Companies already increasing their product portfolios for the home environment include Apple with their mobile handsets, iTunes, apps stores and TVs, and Microsoft, with presence in handsets and TV gaming consoles. Other companies who may be well placed include HP, with their acquisition of Palm, and Samsung, with their strong presence in handsets, TVs and tablets, any one of which has the potential to be used as a user control hub for interacting with the home network.

One particular company focusing a lot of their resources into the mobile home environment is Motorola, which appears to be moving towards marketing its range of cable TV set-top boxes as data centers and "mini-clouds" in order to manage multiple devices and downloading of data. Motorola's vision is for the "mini-cloud" to become the repository for all the information, data records and digital media entertainment a family needs.

Using your mobile handset more frequently in a home environment in the future will create new opportunities for creative handset designers to match the interior home design. In addition, existing home cable TV providers may provide mobile handsets as part of their overall service offering. This is already happening in the UK with Virgin Media, who offer smartphones as well as cable TV and broadband connectivity.

As we look closer into the interaction between the handset of the future and its surrounding network, we will explore the near-field area between the handset and the user and how this is likely to change over the years ahead.

7.6 Near-field sensing

So far in this chapter, we have looked at future trends and their likely impact on materials, user interface and application design of mobile handsets. We now take a closer look at handset industrial design, the sensory environment around the handset and potential form-factor design trends.

The first time we come across the use of near-field sensing in Sarah's journey is when she waves her handset at the ticket vending machine to make a mobile payment. She uses near-field communication (NFC) technology, which is a form of short-range wireless contactless technology embedded in her handset. Today, nearly all the top Tier 1 handset OEMs plan to have NFC integrated handsets in their portfolio, and the Google Android platform Gingerbread 2.3 and onwards can operate with NFC. Some of the key players behind NFC are banking institutions, electronic ticketing vendors and electronic key providers. According to market research firm iSuppli Corp, 13% of all handsets shipped will integrate NFC by 2014.[3]

In the future, there will be different types of near-field sensing technologies available, providing ranges from a few centimeters to a few meters. In the following applications, we will discuss the use of microphones, scent detection sensors, embedded projectors and augmented reality as future enabling technologies that can provide an immersive experience surrounding both the user and the handset, and their likely impacts on the handset form-factor.

In Sarah's journey, she uses a bracelet-shaped handset which is personalized to provide health-related data applications and services via a small flexible display. Flexible e-ink display technology or flexible plastic displays could be used in such bracelet handsets within a few years' time. The potential use of mobile handsets for health-care applications is enormous, and the example of using scents and aromas in mobile handsets is not so far away.

The Philips Research center in Cambridge, UK, is undertaking research into the use of wearable mobile devices and scent monitoring for well-being applications such as reducing stress and sleep disorders. Specifically, using patented technology, a device can be triggered by an electrical signal through a built-in micro-fluidic micro-electro-mechanical-systems (MEMS) device, which can be embedded in "smart" textiles, clothes, jewelry and wearable mobile handsets.

Another innovative approach towards using the space around the handset as a form of communication is being developed by another Cambridge (UK) based company called Intrasonics. Intrasonics has developed a unique encoding platform to enable viewers and listeners to engage

[3] Source: December 20, 2010, http://www.isuppli.com/mobile-and-wireless-communications/news/pages/cell-phone-mobile-payment-market-set-for-take-off.aspx

directly with their favorite media programs and advertising brands by enabling any broadcast media to interact directly with their mobile handsets through data-encoded audio. Specific data can be embedded in a radio station broadcast that a user is listening to. The user runs a pre-installed application on their handset, which listens to the radio broadcast via the microphone and decodes information within the audio. This information could contain details, for example, of the publisher of a book that is being profiled during the radio broadcast, and information decoded from the audio is then displayed on the handset screen. A similar application is being used by Best Buy, a large US retail store, in San Francisco. In Best Buy stores, a US-based startup called "shopkick" has installed small beacons that emit ultrasound signal at a frequency that can be picked up by a handset's microphone but not by human ears. The pre-installed handset application decodes the signal and contacts shopkick's database to work out where the user is and retrieve a Best Buy shopping reward, based on a specific promotion. The opportunity is to use the space around the user to deliver a very personalized experience with appropriate "call to action" moments. Business model innovation may be required in some of these emerging areas in order to monetize the new usage scenarios created. From a handset design perspective, providing a service by using the handset's microphone, a dedicated downloadable application and an encoded audio signal opens up the market, not only to high-end smartphones, but also to mass-market feature phones.

Other areas of near-field sensing described in the story are augmented reality and wireless power charging, which we will describe in more detail below.

7.7 Augmented reality

Augmented reality (AR) is a term referring to a live, direct or indirect view of a physical, real-world environment with elements augmented by virtual computer-generated sensory output, such as graphics. In Sarah's case, when she got off the train before heading into the office, she used her handset to scan her immediate vicinity; she could then see the shops that were local to her environment on the display. This local environment was augmented with the additional layering of descriptive data onto her handset screen, i.e. describing what the shops sell and special deals that they may have on offer. This form of augmented reality in mobile handsets combined with real-time GPS data (to provide a location fix to ensure the correct augmentation is provided) is expected to be used by a growing number of smartphones in the next few years. Already in the market are geo-tagging applications running on handsets, which are being used for mobile social gaming. With advancements in display technology, 3D graphics rendering and with an increasing number of consumer brands having a presence in virtual worlds, we believe augmented reality will enter the mainstream of our daily lives.

According to Gartner, a leading technology research firm, it is estimated that by 2014, 30% of mobile subscribers with data-enabled handsets in mature markets will use augmented reality at least once a week.[4] This will require handset designers to consider how users will interact with their devices differently, when they begin to increasingly view the real world through the lens of the camera on their smartphone.

One of the current drawbacks of using augmented reality is its high power usage, due to the use of GPS, camera and graphics processing, and the need for good illumination of the display. As mobile handsets become ever "smarter," becoming powerful mobile computers, the management and supply of power within the handset is becoming an ever more important issue. One consequence of this is the need to charge devices more frequently, and we therefore believe that we will see the emergence of wireless charging hotspots embedded into our everyday surroundings.

7.8 Wireless power charging

During Sarah's journey, she was able to power up her mobile handset on the train's passenger table, the meeting room table, the coffee lounge table and even the car dashboard resting area. This was achieved through wireless power charging, where power is transferred from a transmitter in

[4] Source: September, 2010, Slide 2, http://www.slideshare.net/HiddenCreative/marketing-with-augmented-reality-an-infographic

a resting mat and into a receiver embedded in the mobile handset which connects to the charging circuits of the handset.

In early 2009, Palm, a former US-based handset manufacturer, launched a handset called the Pre, which could be wirelessly charged by touching a mat called the Touchstone. Later that same year, a company called Powermat launched a range of accessory devices that allowed Apple iPhones and RIM BlackBerry devices to be charged wirelessly. This was achieved by replacing the original back cover of the handset with a cover containing an embedded receiver, or for the iPhone a receiver-embedded sleeve. Handsets are charged by placing them on mats with embedded transmitters, that in turn are plugged into the mains power supply. It was possible to place up to three handsets on a single mat and charge the handsets through magnetic induction.

According to the market research firm In-Stat, the market for wireless charging is set to reach $US4.3 billion in total market revenue by 2014, with iSuppli Corp, another market research firm, predicting that 235 million electronic devices will ship with wireless charging capability by 2014. The majority of these devices will be mobile handsets, although this projection includes notebook PCs, digital cameras and other portable consumer appliances in which this technology can be used.

The adoption of wireless charging can be used as a case study to help understand further how a new technology can enter the market, the various technical and marketing challenges faced, and the impacts on handset design, all of which need to be considered. A key challenge is achieving widespread adoption in the mobile industry, as, once achieved, this drives the economies of scale, which can produce an acceptable price point for mass-market adoption.

7.9 Case study: wireless charging in smart devices

User research suggests that users are interested in the concept of wireless charging, but that it is critical to address a number of factors in order to achieve widespread market acceptance. These include ease of use (e.g. simply placing the handset on a mat), no negative impact on the sleekness of industrial design (e.g. no bulging batteries), and the ability to have a range of handheld products, not just phones, charging simultaneously on the same mat. One resistance to adoption has been the need with some solutions to take off the original back cover of the handset in order to replace it with a modified back cover which is slightly thicker.

The market leader of wireless charging technology for mobile smartphones in 2011 was Powermat, with an estimated 70% market share. Powermat initially entered the market in 2009 with a range of handset accessories. These accessories proved very popular, with over 750000 units being sold in the first two months. The advantage of an accessory is that little or no hardware integration is required with the handset OEM, providing a rapid time to market in which to introduce a new concept. However, in the mid-term a more integrated approach with the mobile handset makes good sense in order to meet the user requirements identified, and thus achieve mass-market adoption.

To embed wireless power charging receiver technology into mobile phones, the electronics must be able to communicate with nearby chipsets and power management electronics, whilst relaying back to the user the status of battery charging, including when charging is completed. To achieve this, requires adding new commands to the handset software to allow the user to view the charging activity. From a handset design perspective, the integration process involves a range of individuals cooperating together including hardware, software and mechanical engineers, as well as user experience designers. In addition, input from industrial designers is needed to minimize any increase in the overall handset size due to the presence of additional electronics. This level of handset integration has many implications for the overall handset design, as well as chipset and operating system integration within the mobile handset. It is this high level of integration and resources needed by handset designers which means it takes considerably longer than might be thought to achieve mass-market take-up of a new technology such as wireless charging.

In the longer term, a better way to integrate the electronics is to embed the power charging capability into a chipset, initially a further custom chip, with opportunities to integrate capability into one of the core chipsets as volumes rise. Some of the largest chipset suppliers in the world, such as Freescale, Texas Instruments, National Semiconductor, Qualcomm and ST-Ericsson, are also now emerging players in the wireless charging industry.

Despite a positive and upward growth for wireless charging in smartphones and other smart devices, there has been a battle between key technology providers around inter-operability and standards, as different organizations vie for position in the market and seek to build on their own strengths. The Wireless Power Consortium (WPC) was formed in 2008 to create a universal wireless power charging standard, so that electronic products and charging areas using the same standard could recognize each other and charge. The intention here is to avoid both the fragmentation of the market and the risk of incompatible solutions – for instance, walking into a coffee shop, resting your mobile handset on the charging surface table, and being told that your handset cannot work on that particular mat because the transmitter mat is made by one vendor and the receiver in the handset is made by another. Members of the WPC include HTC, LG, Samsung, Sony Ericsson, Philips, Nokia, ST-Ericsson, MediaTek, National Semiconductor, TI, Atmel and many other leading companies from the mobile handset industry. An interesting observation is that one of the founding members of the WPC is Fulton Innovation, which is a supplier of wireless charging technology, alongside Duracell and PowerKiss, who offer wireless charging.

It is interesting to observe how transmitter electronics is shaping the environment where wireless charging in mobile handsets can be used. Recently, General Motors announced a $US5 million investment in Powermat. Additionally, Toyota has expressed interest in this area through an announced partnership with wireless charging company 'WiTricity. Office table manufacturers and interior wall furnishers are also looking to integrate wireless charging as "surface" charging. One can imagine people sitting around a coffee table and having their handsets charged on the table surface while they chat. At Consumer Electronics Show (CES) events, there have been demonstrations of the future of surface charging by showing not only smartphones being wirelessly charged, but also kitchen appliances, wall lighting and even cocktail glasses on kitchen surfaces and vertical walls.

The future choice of which wireless charging technology vendor and standard to incorporate into the handset design will not simply come down to technical factors such as the best charging times, efficiency ratings, thickness of electro-mechanics, thermal management and ease of technical integration. In addition, the ability to scale the technology to other consumer devices, such as tablets, and the opportunity to enable the transmission of data (such as real-time video) as part of the power transmission, are important technology roadmap issues. Easier design integration and greater inter-operability will encourage a greater mass adoption of wireless charging and provide an ecosystem of charging devices and charging surfaces in everyday surroundings. This should, for instance, make easier the experience of a group of friends charging their mobile handsets whilst they relax at a table enjoying coffee together.

As with many new technologies, adoption is likely to happen first in the high-end handsets which are less price-sensitive, such as smartphones. It is anticipated, however, that, as the price of wireless charging technology decreases, it will be adopted by the mass market and in the ever more price-sensitive feature phone segments.

7.10 Social values and shared experiences

If we take a snapshot of some of the handsets that have been launched in the early 2010s, we can observe amongst the trends the following:

- an emphasis on application stores and the wide range of choice of apps;
- a strong focus on accessing and updating social networking sites from the handset;
- a growth in the number of handsets with large, high-quality, high-resolution touchscreen displays;
- the ability to multi-task between different apps via the user interface.

Many of the core advertising messages and imagery used around the promotion of these handsets consistently show pictures of a person engaging with others via their handset whilst looking at the handset screen. The interaction, in some instances, could be a video face-to-face call, or

Table 7.1. *Snapshot of handset core advertising messages and descriptions*

Handset brand	Nokia	HTC	Apple	Samsung	Motorola	RIM BlackBerry	Sony Ericsson	LG	Microsoft
Handset model	N8	Wildfire	iPhone 4	Galaxy	Droid X	Torch 9800	Xperia X10 Mini Pro	Town GT350	Dell Venue Pro
User input method	Touchscreen	Touchscreen	Touchscreen	Touchscreen	Touchscreen	Touchscreen plus keyboard	Touchscreen plus keyboard	Touchscreen plus keyboard	Touchscreen plus keyboard
Core advertising message	Shoot HD video in 720p resolution, then stream your footage to your TV in HD with the HDMI-out connection	"Bring your friends to you. See everyone's Facebook, Twitter and Flickr updates, all in one feed"	"This changes everything. Again."	"Android™ brought to life . . . on the world's brightest 4.0″ Super AMOLED screen"	"Every experience from messaging to movies . . . crystal clear calls"	"Discover the new BlackBerry Torch 9800 . . . real-time chats with BBM™"	"Get more than just another fashion accessory with the Xperia™ X10 mini pro in pink. Download all the latest Android™ apps . . . "	"Social animals . . . out and about . . . send an e-mail to friends on your easy touch Qwerty keypad, or post your update on Facebook"	"Designed to get you in and out and back to life," and "Where Microsoft will save you from the confusion of current handsets in the market"

Core advertising image	Handset with video play image	Handset showing time, messages, mails, Internet and camera in the middle of a word showing “You”	Two faces on the front screen engaged in a video call	Handset showing several Google search applications and Google Maps image	Face of a woman calling with arrows describing the user interface and handset features	Handset closed with front screen shown with photos, followed by handset open with sliding Qwerty keyboard exposed	Woman with a pink handset in her hand and next to her images of the handset open and closed showing the Qwerty keyboard	Young people reaching out and touching each other’s hands with an image of the handset showing Qwerty keyboard sliding out	People using the mobile and ignoring everything going on around them

perhaps messaging a friend via Facebook. The important thing missing from these characterizations, however, is the social exclusion from other people in the person's immediate social gathering. There is a lack of participation in a shared experience. Many of the marketing messages and communication is aimed at one person interacting with one handset. This creates an opportunity now and in the future for handsets to have more applications, features and displays which encourage a shared view and immersion of the full handset experience amongst groups of people. In Table 7.1, we show a descriptive summary of a number of handsets launched in the early 2010s. It is interesting to observe the focus of the marketing messages on the individual and their interaction with other people via their device, perhaps at the cost of shared experiences with those who may be in the same room.

We now share some findings on parents' views of mobile handset usage amongst their children, as we explore the impacts on use of discretionary time and shared experiences. In early 2010, Scholastic, a global publishing house, in conjunction with Quinley Research and Harrison Group, conducted a survey examining family attitudes and behaviors regarding reading books for fun.

The key findings of this research, based on a US nationally representative sample of 1045 children aged 6–17 and their parents (2090 total respondents) were as follows.

- Parents believe the use of electronic or digital devices negatively affects the time kids spend reading books (41%), doing physical activity (40%) and engaging with family (33%).
- From ages 6–17, the time children spend reading declines, while the time children spend going online for fun and using a mobile handset to text or talk increases.
- When asked about the one device parents would like their child to stop using for a one- or two-week period, they most often cite television, video game systems and mobile handsets.

One of the parents stated

"As my child becomes more and more involved with electronic or digital devices, I worry that he/she will be less interested in reading books for fun."

A typical scenario today in a European family home could involve a son playing with his Nintendo Wii gaming station, a daughter on her iPhone, a father on his BlackBerry and a mother on Facebook chatting with her friends via her wirelessly connected notebook PC. In this scenario, connectedness within the family unit is being lost because of the presence of multiple smart devices focused on satisfying individualized experiences. We think this is a missed opportunity to use technology to bind people together, especially as the handset is foremost a communications device!

There are companies outside the mobile handset space exploiting this opportunity for shared connectivity in the home context. Examples include Nintendo, Sony and Microsoft with their motion-sensing and dancing-game home entertainment systems. Nintendo's Just Dance 2 sold five million units within three months of its release in October 2010, making it Nintendo's best selling game. This game encourages groups of people to dance to various music tracks together. Microsoft shipped eight million Kinect game system units in the first two months of January 2011. (Kinect is a motion-sensor system for its Xbox 360 video game console, which supports multiple players in the same room engaged in the same game.)

These home entertainment systems can encourage groups of friends and families to share an experience through a single device, and provide a more hopeful view of family life enriched by technology. We believe that further opportunities exist to use the handset to enhance shared experiences when people are physically together, be that within the family or within other social groups.

In the future, we are likely to see some handsets supporting even larger displays and support for 3D gestural interaction, which we believe can encourage multi-person user interaction, strengthening social connectivity through physical proximity. For example, research is being carried out at Ishikawa Komuro Laboratory in Tokyo, where they are demonstrating 3D user interface gesturing and navigation on a handset using the handset camera.

The popularity of smart devices with large displays is very evident today, with one in three US online consumers forecasted to have a tablet

PC by 2015, according to Forrester, a leading US research firm. This rapid growth in demand has the potential to cause problems for mobile phone makers due to global supply shortages of organic light-emitting diode (OLED) displays, widely used in mobile handsets and tablet PCs. In addition, a number of other innovative large-format displays are under development, as described in the following.

Returning to Sarah's journey, we saw her rolling out a display from her handset and reading bedtime stories to her children. This may sound like science fiction to many, but rollable display technology is already beginning to make an appearance.

At the Consumer Electronics Show in Las Vegas, in January 2011, Samsung demonstrated flexible and transparent active matrix OLED displays that can be rolled up like a newspaper. The displays are able to offer the same level of power consumption, color saturation, brightness and contrast as existing active matrix OLED displays in high-end smartphones. Additionally, significant research and development is being carried out at the display technology center of Taiwan's Industrial Technology Research Institute, where it is forecast that rollable displays will be available as early as 2015 for mass-market use. In the meantime, large-format displays on tablet PCs and e-books are providing a readily available electronic reading experience. It is predicted by Forrester Research that 82 million people in the US – one-third of the country's online population – will own tablet computers by 2015. We are likely to see the emergence of e-ink technologies on smartphones over a similar timescale, due to the very low power consumption of the technology.

Technologies also under development that we believe can help address the need for shared proximate experiences include reflective micro-electronic mechanical systems (MEMS)-based displays and pico-projectors.

Qualcomm's Mirasol reflective MEMS display technology is capable of playing full color video with lower power consumption than existing e-ink displays. Furthermore, Mirasol displays require no additional lighting in ambient conditions. This technology could enable groups of people to view videos on smartphones or tablet PCs in bright outdoor environments, overcoming the current constraints of displays which are hard to see in sunlight and which are heavy consumers of power.

An alternative approach to sharing videos and other information with even larger groups of people is through the use of a pico-projector. Pico-projectors are small, handheld versions of larger projectors that one typically finds in classrooms or office presentation rooms. According to Pacific Media Associates (PMA), the global pico-projector market is expected to reach 22 million units by 2014, up from about 700000 units in 2010. We expect the technology to transition from stand-alone pico-projector accessories for smartphones to being fully integrated into the smartphone.

Companies working on pico-projection include Samsung, Texas Instruments, Light Blue Optics and Microvision. At the Mobile World Congress 2010, Samsung demonstrated a pico-projector phone called Beam, and this is likely to be the first of many more handsets to come with embedded pico-projection.

In order to support ever more immersive visual user interactions on handsets, power consumption remains a key design issue. Handset hardware and software platforms are continuing to support more efficient processor technology and improved power management, whilst further reducing the overall bill of material costs, allowing new technology to reach the more price-sensitive, high-volume feature phone segment. In the following section we look further at power consumption trends.

7.11 Efficient handset platforms

Following the handset design process flow from materials through user interaction, industrial and mechanical design, we now look into future trends in hardware and software design, particularly focused around handset platforms.

At the high end of the market, we are seeing growth in multi-core smartphones – with the first dual-core processor based handset announced by LG in December 2010 – the LG Optimus 2X. Multi-core chips provide more processing power without a proportional increase in power consumption, as functions can be run in parallel on different processors, rather than clocking one processor at a much higher speed, which results in higher power consumption. For the end user this translates into faster execution of multi-tasking applications such as real-time social networking updates, gaming, video streaming, web browsing and location-based services. In the future, multi-core chips could enable 3D viewing functionality in handsets without the need for special glasses.

NPD Group reports that the share of handset sales that were smartphones in Q3, 2011 reached 59% for consumers aged 18 and over in the USA. Although the smartphone share is expected to rise rapidly over the next few years, this still leaves a very high percentage of non-smartphones in the marketplace, which will continue to encourage innovative design changes at a platform level for the feature phones. One such technology is known as mobile virtualization.

Mobile virtualization enables several operating systems to run simultaneously on a mobile or connected wireless device. It is a thin layer of software, embedded into the handset, which decouples the operating system, applications and data from the underlying hardware in a secure manner.

Virtualization technology has been widely used for many years in the data server and portable computer industry, and has recently been gaining a foothold in the handset industry due to its ability to make smartphones cheaper and feature phones smarter. Other key drivers for its likely adoption by handset OEMs include a reduction in handset development cycles and the ability to increase program and data integrity by having the OEM's own real-time operating system working alongside existing popular operating systems such as Android or Symbian.

From the perspective of new handset development, having a consistent software platform between the hardware and operating system removes the need to re-design the hardware to support different operating systems, since only the virtualization software platform requires change. The reduction in time to market and reduced development cost greatly improve the OEM's ability to respond to the marketplace more rapidly.

An example of a commercial product utilizing mobile virtualization is the Motorola Evoke QA4, which was launched in 2009. The Evoke is a BREW-based feature phone that is able to run a "smart" Linux-based application operating system. It comes preloaded with Myspace Mobile, Google Quick Search and YouTube.

Major players in the mobile virtualization industry include two US-based software companies, Open Kernel Labs and Red Bend. In September 2010, Red Bend acquired VirtualLogix, a provider of mobile virtualization software. Open Kernel Labs and Red Bend have a combined reach of over one billion handsets using their software solutions.

With mobile handsets becoming ever richer in features, we now look at how the retail channels are likely to capitalize on this in the future and how handset design plays a key role.

7.12 Handsets and the retail experience

If we look back at the story at the beginning of this chapter, we see that Sarah had several handset retailing experiences. One of these was going into a department store and purchasing a handset. Others were related to gathering and sourcing information about a particular consumer purchase and delivering this information onto her handset. We will now take a further look into both these areas of in-store purchasing in handset retail stores and out-of-store retail information gathering via mobile handsets.

In Europe, Carphone Warehouse have transformed their retail stores from small stores with largely undifferentiated handsets, to a much more open environment where customers can move around more freely and see a wide range of "wireless devices" rather than just handsets. Based on an initiative called "Wireless World," Carphone Warehouse plan to broaden their offering to embrace handsets, tablet PCs, notebooks, accessories, specialized content, carrier and broadband connections and technical support. Some of their stores have recruited gaming, computer and notebook specialists to ensure that customer enquiries can be dealt with meaningfully and quickly. This is being supplemented by easier to read, jargon free literature explaining the difference between various handset operating systems and apps stores, and allowing customers to have applications installed on their handsets at the store which are personalized to their tastes, based on discussions with retail staff.

Large supermarket chains are also expressing interest in venturing into the high street with their own handset retail stores. Examples include Tesco, a large UK-owned supermarket chain, whose trial handset retail store based in Bristol, UK, is reported to have out-performed other Tesco in-store retail opportunities.

There are still poor examples of retail environments, which lack the atmosphere of exploration and creativity which customers are increasingly looking for to help them understand the different features and applications on offer. It would be good to have parts of handset retail stores looking like the sitting area of a house or airport business lounge, or even a coffee lounge. In these environments, customers could have the experience of picking up and using a tablet PC, a gaming console, an e-book or smartphone and interchanging the different experiences whilst sharing content with a large-screen TV display on the wall in front of them. This is where handset designers, especially usability designers, have opportunities to work closely with retail management to design the optimum experience for engagement with the mobile handset user.

The handset retailing experience within stores can also be expanded to outside the store – where shoppers are already using their existing handsets to find out more about their next handset purchase before they even step inside. According to a recent survey in the USA by ForeSee, out of 10 000 online shoppers, one-third of the respondents had used their phone to access a retailer website, and an additional 26% indicated that they planned to access retailer websites or mobile apps by phone in the future.

The opportunity exists for existing handsets to provide features to help users choose a replacement handset by allowing comparison of new features, tariffs and contracts online before they enter a retail store. Promotion discounts could also be downloaded before the consumer enters the store. This leaves opportunities for consumers to spend more time in the store to try out their handset and enjoy the retail experience.

7.13 Summary

We have described how the mobile handset is likely to change over the next few years. Key themes emerge from a greater emphasis on environmental footprint, cloud computing, proximity sensing and context, wireless charging and opportunities for a shared immersive user experience around the handsets.

In addition, many new types of device will be created in the future which will also have a wireless connectivity capability. At a keynote presentation on April 4, 2011, Hans Vestberg, CEO of Ericsson, stated that he believed there would be "50 billion connected devices by 2020." He went on to say "this will happen because of three pillars: mobility, broadband and the cloud." The connection of virtually every electronic product, via wireless, to the Internet, in the so-called "Internet of things" could result in transformational change to our homes, hospitals, schools, cities and transportation, making them smarter by harnessing the power of cloud computing. The mobile handset of the future may therefore be co-located with numerous other smart connected devices which will allow the user to connect, share and control many other devices, providing the experiences described in Sarah's story at the beginning of this chapter.

There may be specific technologies we have missed out due to the constraints of the format of this book. In order to continue to provide fresh insights, we refer the interested reader to a website which accompanies this book, where we seek to capture new trends and observations, as well as feedback from readers, regarding further developments in the industry. The website address is www.mobilehandsetdesign.com.

7.14 Conclusion

With the fourth-generation mobile broadband standard approaching, we will continue to see increases in the volume of data traffic being accessed from even more advanced smartphones delivering even richer user experiences. Some handset OEMs have stated "We have direct control of the end user experience." If this were to be part of their strategy, we would encourage them to re-think. Hopefully this book shows that the future is about collaboration and sharing of the user experience. Additionally, we believe it is not just about the user experience of operating the device, but rather it is about the whole value chain and the many touch points. These range from the point of first hearing about the handset, through the retail experience of selecting and purchasing the handset, into its many and varied uses, and finally the process of disposing of the device. Future success depends on strong partnering and a wide participation in the whole value web. Innovation will continue in business models and revenue sharing to reward the many parties involved in delivering the overall experience. The user experience is far too rich and exciting to be owned by just one company. Looking back at Sarah's journey, we can see how various service providers made her experiences more relevant by providing the right coffee, printing services and discounted travel tickets at the right place and at the right time. This demonstrates the power of collaboration and partnerships between service providers, carriers and, of course, the handset OEM, to provide a compelling user experience.

In the future, engaging the emotions of the user and obtaining the balance between human factors and mobile technology will become ever more important. As the technology advances, so too does what people want to do with their devices, and the needs that they seek to have met.

A big question remains regarding whether mobile service providers can manage the growing demand for data through their networks whilst retaining a good user experience at an acceptable price point. Many opportunities exist for partnerships within the content, service and advertising media to support the development of new business models. Advances in handset platforms and their underlying component parts will continue to fuel innovation. Close co-working across the complete value chain from apps developers to handset designers and network operators and technology suppliers is needed to ensure all parties are aligned closely on the technology roadmap as well as the "desirability" roadmap.

Over the years, we have moved from carrier-specified handsets through to OEM-specified handsets, and now to "user-specified" handsets through personalization and the availability of app stores, content and services. The handset industry is now creating mobile social platforms for users to create their own handset experiences. With this thought in mind, we end this chapter by going back to the story of Sarah and how she starts her next day.

Sarah sits at her breakfast table with her daughter's hands clasped around her eyes excitedly, and then a present is placed in front of her. Sarah opens her present to find a beautiful new mobile handset with a special sporty running grip on it and a signature of her daughter's name engraved on the handset and displayed on the welcoming screen. Sarah has never seen such a mobile handset before and so asks her daughter how she managed to obtain this. She replies that she designed the handset online with a group of friends from different countries and their design was voted in the top three by over 100 000 people online. Since they also had reached the target number of group purchases for her designed handset, her social online broker had given the go ahead to have the handsets made and distributed. Sarah smiles as she looks into the eyes of her 11 year old daughter, and together they laugh and play with their new mobile friend.

8 Conclusion

From the first mobile telephone call on a street in Manhattan, New York, in 1973, through to the six billionth mobile phone connection just short of 40 years later, the mobile handset has transformed our ability to communicate and connect. Its level of sophistication is astonishing, yet it has weaved its way so inextricably into the fabric of modern life that it is already challenging to think of how we ever survived without these devices. The next 40 years are sure to be just as exciting a journey, as the handset increasingly becomes our fundamental tool for interacting with people, information and things, both in the real world and the virtual world.

We have taken you on a journey from the early history to the foreseeable future of mobile handset design. Along the way, we have covered all of the key areas of handset architecture and technology design, and uncovered the most important drivers and influences on mobile handset design.

Advances in electronics following the invention of the transistor, at a pace described by Moore's Law, have led to what we might call the "more laws" of: more processing power, more miniaturization, more complexity, more economies of scale, more market growth, more market diversity, more utility value, more apps – and ever more mobile handsets! However, counter-balancing these "more laws" are the unbending physical world constraints of limited spectrum availability, limits on the achievable information transfer through a communications medium and limits on the chemistry of batteries and the resultant battery capacity. In addition, there are practical limits on human cognitive load and manual dexterity which affect our ability to interact successfully with small physical devices that are becoming more and more complex.

In the space between the opportunities created by "more" and the challenges created by the hard limits, innovation and design thrive. Innovation is the fuel of the mobile handset industry. Design – according to the late Steve Jobs, founder of Apple – is the soul. Successful innovation generates new, exciting and often surprising capabilities. Successful design gives us products that thrill us, products that become part of everyday life, products we cannot live without. When technology runs ahead of design, the results are difficult-to-use products that do not survive beyond the early adopters, and do not transform how we use our devices. When technology and design run in tandem, incredible new experiences result which change how we use our handsets, and which may in turn transform the very industry that gave birth to them.

Another theme running through this book has been the emergence of what we have called a "holistic design approach." By this we mean recognizing the inter-connectedness of a wide range of design issues and the evidence of the marketplace that successful handsets are designed by teams of experts, with a clear sense of the over-riding DNA of the handset, and how this connects with the end user at all of the different touch points. A strong collaboration dynamic between technology providers, handset designers, manufacturers, network operators, consumer brands and app and service providers is required to bring successful products to market.

At the heart of an engaging and pleasurable mobile handset user experience is excellent engineering integration between the hardware, software, industrial, mechanical, user interface and manufacturing design – so that, to the user, all of the technical complexity of the handset is invisible to them. Good design is often understated. Aesthetics aside, if the user does not need to think about the design of the product, and rather is able to just enjoy the benefits from using it, then great design has probably been achieved.

There is never time to design a handset completely from the ground up. The market and technology would have changed multiple times in the intervening period, and the economics of doing so would never add up. Consequently, a platform approach to many aspects of handset design is absolutely critical. This is true at the individual hardware and software component level as well as at the handset reference design level. Equally, it is true at the test house, in the factory, through the distribution channel, at the retail outlet and through the app store.

The very high-volume economics of the mobile handset industry have resulted in numerous casualties of former well-known brands that have suffered significant changes in their fortunes. It is a tough business for OEMs, as the margins for error are small around a whole range of issues including core technology, market development, supply chain, distribution, user experience, carrier relationships and differentiation. We believe there will continue to be a small number of successful high-volume manufacturers, with the brand names changing periodically. In addition, there are increasing opportunities for relatively low-volume, high-value brands to be successful in a large range of market niches. This is possible because of the availability of common handset platforms. These platforms can be used by multiple brands using a large number of common components to manufacture products economically. Niche brands may be able to operate with higher margins, so the market conditions are correct for non-traditional handset brands to create strong experiences focused around particular user segments. This could be, for example, a new fashion brand, or even a leading online games publisher with access to large numbers of exclusively developed games for the mobile handset.

So what really *is* a mobile handset and what does the future hold? We see the mobile handset increasingly becoming our interface or "window" into the new world we are constructing online, representing us in this virtual world and allowing us very "analog" human beings to engage fully with each other and with the growing digital library of human knowledge – and trivia – online. Equally, because mobile technology allows us to be wherever we want or need to be, we see the future mobile handset helping to enhance our understanding of our immediate surroundings, including augmentation of information from the online world and interaction with nearby physical objects.

This book has introduced you to many aspects of mobile handset design, and we are extremely excited and passionate about the speed this industry is moving at and the opportunities this creates. We are very optimistic about the future of the mobile handset industry. The industry continues to undergo rapid change. We believe the best innovative ideas will come from the dreamers, who see the possibilities rather than the constraints.

Appendix
User interaction and experience design phases

There are several design phases that contribute to the user interaction (UI) and experience. In the following, we present a typical design-phase list, which outlines the steps followed in the design process.

1 User interaction and experience design phases

(1) Requirement list.

(2) Requirement validation:
- analysing and validating the UI requirements list with the UI specification team and software developers;
- user validation of the feature list;
- background study;
- recording of user requirements.

(3) UI logic:
- identifying cases of essential use;
- overall UI logic specification;
- overall UI structure;
- documentation and negotiating contents of documents with software development team;
- UI logic in each individual module;
- application flow design and features definition;
- documentation of use.

(4) Overall UI design:
- UI concept of the overall interaction modes and metaphors;
- UI specification for input to the industrial design;
- graphical user interface style and concept guide;
- UI input and output methods of development, specification and prototyping.

(5) Sound design:
- designing the UI audio feedback.

(6) Graphical user interface (GUI) layout design and specification:
designing the navigation logic;
menu systems and selection;
exiting and saving logic;
GUI layout wireframes;
individual applications and physical applications, including
- contacts and address book;
- dialing and call mangement;
- text input;
- messaging;
- camera.

(7) GUI layout:
GUI components design;
GUI transition animations;
animated icons.

(8) Flash demos:
creating a flash demo of the overall UI concept.

(9) Usability studies.

Glossary

air interface	the radio communications link between a device such as a mobile handset and a base station.
Android	popular open source operating system for mobile handsets developed by the Open Handset Alliance led by Google.
augmented reality	refers to the overlay of additional information providing sensory input to a user which adds to or augments the user's perception of their current physical real-world environment. Most current augmented reality solutions augment additional visual input, for example providing an overlay of information related to the physical world on a camera.
bandwidth	the range of consecutive frequencies available for transmission and reception of electromagnetic signals on a given transmission medium, measured in hertz (Hz).
Bluetooth	wireless technology standard for exchanging data over short distances.
brand DNA	represents the core values of a brand, which good design seeks to encapsulate throughout the experience of using a product or service.
cloud computing	an approach to providing computing resources as a utility service, on demand, over the Internet, using large numbers of servers.

design language	a brand's visual expression that guides the design of a complement of products or architectural settings.
feature phone	mid-range phone positioned to achieve many of the capabilities of a smartphone for a lower cost point and with lower performance.
GPS **(global positioning system)**	navigation system based on triangulating location from a constellation of satellites in low Earth orbit, about 12000 miles above the Earth's surface.
IM **(instant messaging)**	a form of real-time messaging operated over the Internet. IM is traditionally text-based, though any number of media formats may be supported, most notably video.
ODM **(original design manufacturer)**	the company who designs and manufactures products to a customer's requirements. Customers are typically traditional handset OEMs (q.v.), though they may also be other organizations such as network operators, fashion brands, retail brands or essentially any company prepared to brand and take responsibility for the product.
OEM **(original equipment manufacturer)**	the company who brands and markets a product, and is responsible for all aspects of the product from specification, design and manufacture, through to sales, marketing, service and warranty. Originally, OEMs were fully horizontal integrated organizations, undertaking all aspects of the product lifecycle in-house. Many OEMs now outsource a significant proportion of their design and manufacture to third parties, such as ODMs (q.v.), whilst working with brand and design

	consultancies to create and manage the overall product proposition.
open source software (**OSS**)	software made available in source code format, under the terms of an open source license. Such a license typically allows the software to be improved and redistributed without a fee being due to the licence holder, in return for contributing back, under the same licence terms, any source code improvements made to the software.
OS **(operating system)**	the software infrastructure of a computing device which manages the allocation of resources such as processing time, memory and access to hardware peripherals on behalf of software processes or tasks which perform the functions of the software system.
PCB **(printed circuit board)**	used to support electronic components mechanically and to connect them together electrically.
rendering	computer image of a model containing geometry, viewpoint, texture, lighting and shading information.
roadmap	time- versus capability-based plan, indicating the intended or likely future direction of technology, products or services.
smartphone	high-performance mobile handset with Internet connectivity and an advanced operating system such as Android, iOS, Windows Phone or Symbian.
ultra-low-cost (ULC) handset	a handset with basic features, which has an over-riding design parameter of minimum cost (typically below US$25).

USB (**universal serial bus**)	industry standard for communication, power and connectivity between computing devices and peripherals.
WiFi (**wireless fidelity**)	formally known as the IEEE 802.11 series of specification, WiFi is a wireless, local-area, high-bandwidth standard for data transfer. It is typically used to provide wireless connectivity between devices and a local access point, which is in turn connected to the Internet or other communications network.

Index

1G, 8
1xRTT, 10, 24
2G, 8, 9, 10, 16, 17, 18, 19, 20, 21, 23, 62, 77, 78, 101, 102
2.5G, 8, 24
2.75G, 8
3D design, 70, 181
3G, 8, 11, 24, 30, 31, 34, 70, 99, 102
3.5G, 9, 11
3.75G, 9
3.9G, 9, 11
4G, 9, 11, 31, 102

Aesir Copenhagen, 172
AGC, 15
Amazon, 34, 205
AMPS, 7, 10, 14, 15, 16, 33, 101
Android, 34, 136, 139, 140, 142, 143, 144, 156, 158, 231
Angry Birds, 48, 51
antenna, 70, 100, 106, 107, 169, 182, 187
API (application programming interface), 136, 139, 157, 160
Apple, 29, 30, 40, 44, 49, 50, 67, 156, 158, 177, 184, 190
application framework, 135, 138, 139, 140
application libraries, 139
application platform, 79, 137, 142
application processor, 72, 73, 82, 113, 123, 124, 125, 126, 129
application software, 72, 78, 130, 131, 134, 140, 161
architecture, 57
ARM, 73
ARP (Autoradiopuhelin), 5
AT&T, 1, 2, 5, 6, 33
augmented reality, 208, 209

Bahrain, 6
Bang & Olufsen, 172
baseband chipset, 72, 73, 81, 113, 117, 154
Bell Labs, 4
Berners-Lee, Tim, 22
BlackBerry, 26, 141, 158, 178
block encoding, 104
Bluetooth, 99, 107, 231
BoM (bill of materials), 59, 60, 66, 125, 170, 183
branch prediction, 114
brand DNA, 85, 171, 172, 173, 190, 191, 193
Brazil, 38, 65
BREW, 29, 34, 157, 220

Carphone Warehouse, 167, 203, 221
carrier signal, 95, 97
CDMA, 11, 14, 15, 16, 18, 69, 102
CDMA2000, 10, 11, 24, 30, 102
cellular modem, 91, 117, 118, 119, 120, 121
CEPT, 12, 33
channel access, 13, 14, 31, 100
China, 10, 30, 186
chip package, 73, 75, 76, 111, 115, 116
chipset architecture, 82, 112, 113, 117, 126
cHTML, 27
clock speed, 25, 112
cloud computing, 53, 179, 184, 198, 203, 204, 205, 206
codec, 113, 124, 125, 126, 140
connection management, 151
contextual awareness, 175, 201, 203
convolutional encoding, 104
Cooper, Martin, 2, 7, 32
CPU, 111, 113, 114, 115
CTIA, 13, 14

D-AMPS, 14, 16
Dalvik, 143
data synchronization, 204
Denmark, 6, 32
desirability, 40, 41, 162, 172
device driver, 78, 81, 135, 142, 143, 144
DoCoMo, 18, 27, 28, 33
DSP, 9, 72, 81, 82, 120, 154
DynaTAC, 2, 6, 7, 33, 130, 132

EDGE, 11, 97
environmental design, 188, 199

Ericsson, 2, 5, 6, 19, 223
ETSI, 13, 33
Europe, 5, 10, 12, 19, 30
European Union, 12
execution environment, 157, 159, 160

Facebook, 54
FCC, 2, 4, 6
FDMA, 13, 101
feature phone, 29, 82, 113, 136, 156, 159, 220
FEC (forward error correction), 103, 104
finite-state machine, 151, 152
Finland, 5, 6, 16, 18, 32
frequency hopping, 98

Germany, 34
Goldvish, 48
Google, 50, 142, 156, 158, 186
GPRS, 10, 24, 25, 28
GPS, 16
GPU (graphics processor unit), 73, 127
GSM, 10, 13, 16, 17, 18, 20, 21, 24, 33, 68, 74, 87, 97, 99, 101, 150, 152, 154

HAL (hardware adaptation layer), 77
hardware design, 70, 163, 166, 168, 169, 183, 190
hardware platform, 47, 77, 163, 166, 169
holistic design, 162, 163, 172, 193, 227
Hong Kong, 16
HSDPA, 11
HSUPA, 11
HTC, 34, 49, 54, 176
HTML, 22, 25, 27, 141
Hutchinson, 16

iMode, 27, 28, 29
India, 51, 54, 65, 198
industrial design, 7, 41, 107, 165, 177, 178, 179, 180, 190, 206, 210, 211
INQ, 168, 177
Intel, 57, 110, 196
intellectual property, 63, 64
interaction design, 176, 179, 190
interleaving, 103
interrupt service routine, 143, 144, 155
iOS, 50, 158
iPad, 34, 51, 69, 158
iPhone, 29, 30, 34, 44, 65, 130, 133, 156, 158, 176, 187, 190
IS-54, 14
ITRS (International Technology Roadmap for Semiconductors), 110
ITU (International Telecommunications Union), 11, 31

Japan, 10, 18, 19, 27, 28, 29, 30, 32, 33, 88, 156
Java, 29, 63, 141, 142, 156, 158

kernel, 142, 143
Kindle, 34, 51, 205
Korea Mobile Telecommunications Company, 7

layer 1, 114, 148, 149
LG, 8, 19, 39, 53, 190, 219
Linux, 141, 142, 143, 158, 220
logical channel, 99, 151
LTE, 11, 31, 34, 87, 102, 107, 110
LTE-Advanced, 11, 31

manufacturing process, 83, 166, 186
MCM (multi-chip module), 116
mechanical design, 60, 163, 165, 166, 180, 182
MEMS (micro-electro mechanical systems), 207, 218
MENS club, 19
Mexico, 6
microprocessor, 72, 82, 120, 143
Microsoft, 50, 156, 158, 179, 186, 204
middleware, 139
MIMO (multiple input and multiple output), 107
mobile network, 8, 68, 74
mobility management, 151
Mobitex, 26
modem chipset, 73, 82, 91, 117, 118, 119, 120, 121
modulation, 95, 96, 97, 99
Moore's Law, 19, 109, 129, 161, 226
Motorola, 1, 5, 7, 19, 21, 29, 33, 34, 39, 132, 141, 156, 186, 200, 206
MTA, 5
MTB, 5
multi-band, 104
multi-core, 111, 112, 219
multi-mode, 105
MVNO (mobile virtual network operator), 48

near–far field effect, 15
NFC (near-field communication), 206

NMT (Nordic Mobile Telephone System), 6, 8, 10, 32
Nokia, 6, 7, 8, 17, 19, 20, 27, 33, 38, 39, 51, 53, 84, 132, 141, 156, 158, 170, 171, 179, 186, 199
Norway, 6, 18, 32
Nvidia, 73

O2, 200
ODM, 48, 85, 167, 168, 186
OEM, 59, 167, 186, 201, 223
OFDMA (orthogonal frequency division multiple access), 31, 102
open source, 63, 141, 142
operating system, 44, 50, 79, 80, 82, 133, 141, 142, 156, 157, 158, 159, 161, 166, 181, 184, 185, 190, 204, 205, 220
OSI (Open Systems Interconnection), 147

PA (power amplifier), 74, 105
path loss, 100
PCB, 61, 70, 72, 74, 75, 83, 85, 165
personal digital assistant, 26, 189
personas, 43
phase gate analysis, 169
Philippines, 18
PHS, 10
physical channel, 14, 99, 101
physical layer software, 72, 80, 81, 130, 153, 154, 155, 161
pico-projector, 191, 218, 219
platform approach, 57, 59, 84, 166, 219
power consumption, 109, 112, 127, 161
power management, 112, 127, 128, 211
Powermat, 210, 211, 212
process geometry, 110
project planning, 166
protocol, 21, 26, 67, 68, 145, 146
protocol stack, 19, 25, 72, 80, 81, 82, 130, 145, 146, 147, 148, 150, 151
prototype, 2, 6, 163, 165, 166, 173, 181, 182, 191

Qualcomm, 14, 15, 16, 29, 183, 218

radio chipset, 72, 74, 75, 96, 115
radio resource management, 100, 151
radio waves, 95, 100, 107
Radiolinja, 16, 18
reference design, 54, 92, 183, 184, 227
RF band, 6, 12, 26, 74, 93, 95, 99, 103, 105
RIM, 26, 141, 158, 168, 178
Rovio, 48, 51
RTOS (real-time operating system), 142, 144

Samsung, 8, 19, 39, 53, 172, 200, 204, 218
Saudi Arabia, 6
SDR (software-defined radio), 117
seven-layer model, 147
Shannon limit, 104
Sharp, 22, 27, 34, 133
Siemens, 19
SIM, 68, 69, 70, 152
SIM application toolkit, 152
situational relevancy, 44
SMS, 10, 17, 18, 33, 147
software architecture, 78, 80, 82, 138, 139, 160
software platform, 78, 84, 85, 184, 186, 220
Sony Ericsson, 34, 188, 200
South Korea, 7, 10, 19, 34, 157
spectrum, 6, 8, 9, 12, 14, 16, 30, 74, 80, 93, 94, 95, 99, 105, 202
spectrum auction, 30, 34
spread spectrum, 14, 99
Sweden, 5, 6, 18, 32
Symbian, 156, 158
system in package (SIP), 111, 116
system on chip (SoC), 111

TACS (total access communications system), 6, 10
Tag Heuer, 48
Tandy Mobira Corporation, 7
TD-CDMA, 30
TDMA, 13, 14, 16, 24, 101
teardown, 58, 59
Tesco, 48, 221
third-party software, 29, 61, 62, 136, 137, 138, 140, 156
Three, 34
T-Mobile USA, 156
Tokyo, 6, 217
touchscreen, 188, 189, 190
transistor, 4, 5, 109, 112, 130
transmit power, 3, 10, 15, 16, 95, 97, 98
trend analysis, 37, 38
Trexta, 186
turbo coding, 104

UK, 6, 10, 34
UMTS, 30, 70
USA, 6, 7, 12, 13, 222

user experience, 9, 25, 43, 174, 176, 204, 223, 224, 227, 229
user interface, 17, 20, 30, 45, 78, 132, 163, 172, 174, 175, 176, 179, 185, 193, 202, 203
user journey, 41, 42, 43

Verizon Wireless, 31, 67
Vertu, 40
virtualization, 220
visual design, 173, 190
Viterbi algorithm, 104
Vodafone, 6, 27, 28, 33, 34, 204

WAP, 26, 27, 28, 33, 62
WCDMA, 11, 68, 102
WiMax, 11, 31
Windows Phone, 156, 158, 204
wireless charging, 209, 210, 211, 212, 213
Wireless Power Consortium, 212
WirelessMAN-Advanced, 11, 31
World Wide Web, 8, 21, 22

Printed in the USA
CPSIA information can be obtained
at www.ICGtesting.com
LVHW010031060923
757354LV00004B/14